We at the Center of the Universe

Other works of interest from St. Augustine's Press

John Lukacs, *Confessions of an Original Sinner*
Patrick Deneen, *Conserving America?*
Peter Augustine Lawler, *American Heresies and Higher Education*
Peter Augustine Lawler, *Homeless and at Home in America*
Peter Augustine Lawler, *Allergic to Crazy*
Albert Camus, *Christian Metaphysics and Neoplatonism*
Pierre Manent, *Beyond Radical Secularism*
Pierre Manent, *Seeing Things Politically*
Rémi Brague, *On the God of the Christians*
(and on one or two others)
Rémi Brague, *Eccentric Culture: A Theory of Western Civilization*
Edward Feser, *The Last Superstition:*
A Refutation of the New Atheism
H.S. Gerdil, *The Anti-Emile: Reflections on the Theory and Practice*
of Education against the Principles of Rousseau
Gerhard Niemeyer, *The Loss and Recovery of Truth*
James V. Schall, *The Regensburg Lecture*
James V. Schall, *The Modern Age*
Josef Kleutgen, S.J., *Pre-Modern Philosophy Defended*
Marc D. Guerra, *Liberating Logos:*
Pope Benedict XVI's September Speeches
Peter Kreeft, *Summa Philosophica*
Peter Kreeft, *Socrates' Children* (in four volumes)
Ancient, Medieval, Modern, and *Contemporary*
Peter Kreeft, *Socratic Logic*
Ellis Sandoz, *Give Me Liberty:*
Studies on Constitutionalism and Philosophy
Roger Kimball, *The Fortunes of Permanence:*
Culture and Anarchy in an Age of Amnesia
Stanley Rosen, *Essays in Philosophy* (2 vols., *Ancient* and *Modern*)
Roger Scruton, *The Meaning of Conservatism*
René Girard, *The Theater of Envy: William Shakespeare*
Joseph Cropsey, *On Humanity's Intensive Introspection*

We at the
Center of the Universe

John Lukacs

ST. AUGUSTINE'S PRESS
South Bend, Indiana

Manufactured in the United States of America.

1 2 3 4 5 6   23 22 21 20 19 18 17 16

Library of Congress Cataloging in Publication Data
Lukacs, John, 1924– author.
We at the center of the universe
John Lukacs.
1st [edition].
South Bend, Indiana: St. Augustine's Press, 2016.
Includes index.
LCCN 2016012557
ISBN 9781587319099 (clothbound: alk. paper)
LCSH: Knowledge, Theory of.
Philosophical anthropology. | Cosmogony.
LCC BD161 .L85 2016 | DDC 191—dc23 LC record
available at http://lccn.loc.gov/2016012557

∞ The paper used in this publication meets the minimum require-
ments of the American National Standard for Information Sciences
– Permanence of Paper for Printed Materials, ANSI Z39.48-1984.

St. Augustine's Press
www.staugustine.net

Table of Contents

At the Center of the Universe

WE ARE AT THE CENTER OF THE UNIVERSE. We ought to recognize this now, at what seems to be the beginning of a new age, for which *postmodern* is an inadequate word. A great change after about 500 years.

There are two (and for some of us, three) reasons why we should acknowledge our centrality. They have inspired and compelled me, after considerable hesitation, to write these words.

....

WITH THIS RECOGNITION I KNOW THAT, although I remain in a small minority, I am no longer entirely alone. I am speaking of the uneasy realization that the so-called scientific view of the universe is insufficient. It is not enough to make this argument on moral or religious or metaphysical convictions alone. There are ample evidences that the scientific or materialistic or deterministic explanations for the world that we know are at best incomplete or at worst insubstantial. The achievements of science during recent centuries have been immense, of course. But in the past 200 years, more and more people, including scientists, have come to the conclusion that the science pertaining to the subjects of their knowledge is imperfect, and probably inevitably so. (Note, too, that the word *science* has narrowed to mean the science of nature; the word *scientist* became current in English only after about 1840.)

We ought to recognize that one of the main applications

of material science, technology, depends on a single limited function of causality, what we call mechanical causality, which Aristotle called efficient cause and defined as but one of four causalities. (The other three are material cause, formal cause, and final cause.) Mechanical causality means that the same causes must necessarily result in the same effects. That is the inevitable condition of machines—while at the same time it is incomplete, because it excludes the presence or participation of any kind of nonmaterial or nonmechanical element. A recent statement by the poet Wendell Berry is most appropriate here: he wrote that in the near future the great division of mankind may be between people who think of themselves as machines and people who think of themselves as creatures. His statement brings me to my argument that the earth is at the center of the universe.

. . . .

ABOUT FIVE OR SIX CENTURIES AGO, at the beginning of what is still called the modern age, Copernicus and then Galileo discovered and proved that the center of the universe was not the earth. Centuries later astronomers' observations have led people to understand that the entire solar system is but a minuscule portion of the numberless galaxies of the universe. But my theme is not cosmology.

Some theologians and churchmen at the time of Galileo's trial for heresy did want to condemn him (as it turned out, the condemnation he received did Galileo little harm.) They offered him a way out—to accept that his treatise on the heavenly bodies was not an attempt to state a definite truth but a hypothesis (which, in a way, it was). Galileo would not agree to that. He believed and said that he had discovered a new truth, a forever truth.

Yet what is given to human beings is not the pure truth but the pursuit of truth. The shortcoming of Galileo and Descartes and Newton and an endless line of scientists and "realists" has been their lack of recognition—or even admission—that all of our knowledge is inevitably only human, with actual limitations —and near-miraculous potential richness, too. They looked at the universe as being *outside* of us and saw the greatest achievement of the human mind as being to discover and prove the existence of more and more matters in a world beyond our existence. But we, inhabitants of this earth, are inside the universe and indeed at the center of it. What we know of the universe is never outside of us.

Perhaps this is how (how, rather than why) reading the great humanists of the early modern age—Machiavelli or Shakespeare, Montaigne, or Pascal—tells us things that may be more enduring than the statements of the scientists of their time. Three hundred years ago, the philosopher Giambattista Vico insisted on the difference between *certum* and *verum*, between the measurably accurate knowledge of nature and the knowledge of our kind. He understood that the human causalities and relationships are more complicated than mechanical ones, since the human mind interferes with and disturbs the relationships of material causes and effects. Human relations and human communications are incomplete—while not at all meaningless. Only consider our inclinations and relations to each other.

The strongest evidence that we are not possessors of eternal truths is the inevitable relationship between the human knower and what he knows—never identical but inseparable. That relationship between the knower and what he knows, the seer and what he sees, the hearer and what he hears, and so forth, suggests that our perception is a component of our reality. A sometimes conscious and at other times less conscious recognition of this condition has grown over the past 200 years. We can find it in Goethe's treatise on color (dismissed for a long time even by his admirers), which says that color has three components: a material-chemical substance, contrast, and the act of seeing. The profound recognition that our reality is not separable from us was beginning to grow.

Here and there among French painters, writers, and thinkers toward the end of the nineteenth century, an impatient reaction to "realism" began to appear—whether or not they were fully aware of what their impatience meant. Consider the "Impressionist" painters. They replaced realism, which they knew had run its course. The scenes and people they painted were not less (but in a way more) real—their "subjects" were what the painters saw, not the places and people by themselves. In their renditions, their participation as viewers (including the technical incompleteness of viewing) amounted to a main component of reality. Human knowledge, including human sight, is participant. (Were they aware of what this means? Perhaps not. But this did not compromise or reduce the quality of their art.)

When it came to prose, the best novelists knew that the novel was different from the epic: they had to see and describe people as they were at a certain time and a certain place. That involved more than the choice of new subjects; it involved a shift in perception and in style. Flaubert himself (often described as a high master of realism) hinted even in *Madame Bovary* that realism was not quite what he intended to depict. His famous principle of the *mot juste* was his insistence that the right word mattered more than the right "fact." Maupassant puts this very well: that the aim of the realistic novelist "is not to tell a story to amuse us or appeal to our feelings; but to compel us to reflect, and to understand the darker and deeper meaning of events"—a desideratum much closer to history than to science.

His observations suggest something contrary even to what Lord Acton, the great liberal Catholic historian, thought throughout his career. Near the end of the nineteenth century, he wrote that the science of history had reached a level where a history of the Battle of Waterloo could be written that would be fully acceptable to French, British, and Prussian historians, since it would be invariably fixed. According to him, professional history had reached a perfect or near-perfect level where it could be regarded as a science. About this Acton was wrong. Science was and is a part of history, not the other way around.

In the twentieth century (a transitional one), a definite proof of the failure of science to include the human element, its involvement by participation, came from great physicists—Heisenberg and Bohr in the 1920s. This is the uncertainty or indeterminacy principle that came out of their researches. They found that our knowledge of atomic particles is the unavoidable result of our methods of observing them, of human participation, of its interference with mere physical causality (uncertainty means we cannot physically define the position and velocity of an atomic particle at the same time). This recognition of inevitability of human participation makes the still widely accepted categories of "objective" and "subjective" knowledge incomplete. Our knowledge of this universe is inevitably participant.

The evidence—and the essence—of this truth is simpler than it may seem. It is the simultaneous and inseparable experience and function of perception and imagination. We see and imagine at the same time. The recognition of this condition of human knowledge accords entirely not only with what I wrote about the knower and the known, but also with the recognition that our world is at the center of the universe. The English thinker and writer Owen Barfield (1898–1997) ascribed some of this to the evolution of poetic diction. More essential is what

> We half-perceive (that is, receive through sense-impressions) and half-create . . . not only the fictions of poets, but also the ordinary physical world. . . . The highest reaches of the imagination are of a piece with the simplest act of perception. . . . Only by imagination . . . can the world be known. And what is needed is, not only that large and larger telescopes and more and more sensitive calipers should be constructed, but that the human mind should become increasingly aware of its creative activity.

. . . .

THE KNOWN, VISIBLE, AND MEASURABLE CONDITIONS of the universe came not before our existence and consciousness, but as a result of them. Our universe is as it is because at its center exist conscious and participant people who can see it, explore it, study it. Such an insistence on our centrality, and on the implicit uniqueness of human beings and of their earth, is a statement not of arrogance but of its very contrary, of humility—a recognition of the inevitable limitations of mankind. Arrogance is the moral limit of those who state that the scientific and mathematical formulas worked out by frail and mortal human beings over the short span of 500 or 600 years—water is H_2O, light travels at 186,282 mps, $E=mc^2$—are absolute and eternal truths, in every place and time in the universe, trillions of years ago as well as trillions of years in the future, just as shortsightedness is the mental limit of those who believe that mathematics and geometry preceded the creation of this world and will remain true even when our world will have ceased to exist.

I appeal to the common sense of the reader. When I, a frail and fallible man, say that every morning the sun comes up here and in the evening goes down there, I am not lying. I do not say that a modern astronomer, dismissing this observation for being geocentric, is lying. There is accuracy—determinable, provable precision—in his observations and assertions, considering his particular measurements of the sun and earth. But my commonsense experience of the earth is more basic than are his formulas. The evidence for the latter may well be true, but not true enough. When I see and recognize a tree in my garden it is inseparable from what I think *and* what I have thought and even what I have imagined about it.

. . . .

THE RECOGNITION THAT THE HUMAN OBSERVER cannot be separated from what he observes and knows suggests that we, and the earth on which we live, are back at the center of the universe—an anthropocentric and geocentric one. But this is both more and less than the returning movement of a pendulum. The pendulum of history and our knowledge of the world never swings back. First came nature, and then came man, and then the science of nature. And now, in the twenty-first century, or perhaps later, we may consider yet another stage of our consciousness and of our place and time in the universe, that of our interiority: that the conditions of our knowledge and our necessarily incomplete understanding of human nature (like all human understanding, even its impulse of love) may be more important than our science of nature. In our complicated democratic age, it seems to me that something like this may happen through a further skepticism about materialism—or, alas, through some oceanic movement of popular mysticism. But my theme is history, not prophecy.

....

ABOUT FAITH, BELIEF, RELIGION at the end of a great age and at the beginning of a new one, I am still bound to speculate. One last reason—or, better, argument—for placing humanity and its earth at the center of the universe: I happen to believe in God, and that Christ *was* his son. (Why I believe this, or perhaps why I wish to believe it, is not easy to tell, being part and parcel of my interior life—something that does not belong here.) Still, what this belief means, and what it ought to mean, is a recognition that Christ's life among us, on this earth, may have been the central event in the history of mankind. If so, then this historical event took place in what was then (and not only then but since and in the future) the center of the universe. I know that, being such a believer, I am among a minority of human beings. And this essay is not especially directed to members of that minority.

To this I wish to add my anxiety about many believing Christians whose belief in Christ may be honest, sincere, and profound. Evidence suggests that their view of the world and of its history now exists together with, or at least alongside, their belief in endless progress, including the power of humankind to know and rule more and more of the universe, beyond this small planet where God makes us live. Sometimes I fear that as the life of Christ—only 2,000 years ago, a tiny portion of what we know of the history of mankind—becomes further and further away because of the passage of time, the meaning of his words, his life, his calvary may weaken in the imagination of men.

But that I cannot know.

BEYOND THE END OF AN AGE

That "The Modern Age" has been a misnomer ought to be more than obvious now. The "modern" adjective appeared in English about five hundred years ago, but "belonging to a comparatively recent period in the history of the world" only in 1823 (O.E.D.). The now current designation of "post-modern," too, is an imprecise misnomer. Of course people during and well after the Middle Ages did not know such terms. But the subject of this essay is not one of the historical definitions. Still their prevalence is a most important development of recent centuries, that of historical consciousness.

The protean and profound development of this involves more than evident changes in the material conditions of our lives, and also of our thinking. It touches even more than our conscious thoughts but our very hearing and seeing. That is what I dare to suggest, to sketch now with trembling hands in the ninety-third year of my life.

....

Hearing, listening, music, including its human creators and audiences, have always existed and will go on to exist. Their conditions are more than the histories of their creators. (Neither is the history of art identical with the history of artists, even when some of them may have been "the antennae of a race.") A little more than one hundred years ago E. v.Cyon, scientist and philosopher and a genius, wrote that the ear is *THE* organ of spatial and temporal sense: "the most important of all of our sense organs." (In 1980 no less a profound thinker than George Kennan was thinking of writing a biography of Cyon, though not about the above subject.) Cyon was one of the first extraordinary thinkers in a variety of fields who around 1900 proceeded to question and at times to demolish the materialist and determinist asseverations and beliefs of "scientific" categories. Still my present subject is not the history of science but something more complicated than that.

The history of music is of course a vast subject, also because of its inseparability not only from its composers and performers but from its listeners. As many other matters near or at the beginning of "the Modern Age" there came a great change about five or six centuries ago. The structure of music gained more dimensions (something for which the term "symphonic" is probably inadequate but I cannot find another one). This had something to do with new and improved instruments, but there was much more than that. The compositions of Bach and later, say, of Mozart or Beethoven included new sounds, new sequences, new sonorities, combined effects enchanting large audiences. The history of music was now inseparable from that of its consumption. This is why this kind of music was prototypical and a classic example of what I, for one, prefer to call the Bourgeois (rather than the Modern) Age. Some of its provenance had something to do with aristocratic courts and circles; but essentially its great success had much to do with its admirers and devotees and consumers in the upper bourgeoisie. Think only of the cult of operas and of opera houses! But my subject is not sociology. In this sense it is well-nigh impossible to reconstruct or even to analyze what had made great and influential audiences centuries ago enraptured by the sonorities and melodies of Mozart or even of Wagner. (There was an element in this where the adjectives "emotional" or "intellectual" are inadequate; perhaps the somewhat cynical German word "Schwärmerei"—sentimental public enthusiasm of a kind—had something to do with it.)

The devolution of music, very much including popular music, in the twentieth century, was prototypical of the ending of an age. Music has not come to an end, it never will be, but the deterioration of symphonic and sophisticated music, with its fairly elegant harmonies, is more than a rapidly passing phenomenon. (With all of their then still extant dependencies of a then recent past, composers such as a Gershwin or even painters such as Cézanne were passing passers, limners of insufficient bridges between what had been a fine past and what they may have thought was to be a promising present.) Oh, yes, rock-and-roll will not prevail, not only beyond Beethoven but also not all beyond a Debussy or a Ravel. But the essence is this: people—including myself—do not need music in ways we used to. We like music, but we don't desire it like so many of our ancestors did

Some kind of new, beautiful and unearthly music will no doubt appear and touch men and women again. But they will not be novel versions of the great bourgeois classics—just as a new Dante or Giotto will not appear again: for not only does history not repeat itself but the history of music, too, is more than hearing: it is listening.

. . . .

It is high time for us to recognize that this kind of devolution involves not only listening but seeing. Indeed we must know that perception and imagination are not only related and not only to some extent dependent on each other. They are *SIMULTANEOUS*. Different profound thinkers in the twentieth century (any list of them would be necessarily incomplete) knew or sensed this. Again their anti-determinism involved more than the history of art and of artists (perhaps an early symptom of this reaction was that of the "impressionist" painters—another imprecise term—whose reaction to "realism" was their wish to depict what and how people were seeing). "Abstract" painting thereafter was a temporary and often meaningless phenomenon. And of course this has gone beneath and beyond any history of "art." But I am concerned with something much more prevalent, which is the present domination of pictures, involving us every day and night. What photography and then television, etc. brought about has been a deterioration of imagination and of perception—*A NUMBING* of the very attention, the practice of seeing.

. . . .

This involves the very relationships of human beings with each other. Not only the history of literature but of people hearing and speaking and listening to each other. It includes more than a decline of conversation (a good word, that). The homogenization of societies has gone together with the decline of neighborliness—of people paying sufficient attention to others. Evidences of such devolutions are endless. They are there in the increasing imperfections of their habits of communications, ranging from the declining practices of writing to people, to addressing each other with standard, often abstract and incomplete formulas and phrases—all of this at the time of universal telephony. "The Information Age!" "The Age of Communications!" The opposite is rather true. "Information" and "Communications" are imprecise words. Much of "in-formation" does not in-form people, while "communications" have nothing to do with community: they are only a process.

That too will change. A time will come when more and more men and women will discover their needs for closer and better relationships. They may experience a yearning for words. This may or may not happen after some monstrous technological catastrophe in the lives of their survivors. Or not. But more and more men and women will seek connections with their neighbors. Will there be new Shakespeares? How can we know? But the need of some people to speak to each other will perhaps be more precious than ever. Before that the substitution of words, by different ways of expressions, including pictorialization, suggests that men are tired of being men, and that women are weary of being women—which is, alas a deep predicament of our times. In the mental order, Simone Weill once wrote "the virtue of humility is nothing more or less than the power of attention"—to which allow me to add that attention has nothing to do with the "subconscious": nothing.

When people understand the limits of their knowledge of the world they sense that their knowledge of these limits does not impoverish them but, perhaps paradoxically, enriches their minds, words, not pictures. Consider the tremendous sentence: "Lord, I am not worthy that you should enter under my roof: but say the word and my soul will be healed." In this respect Pascal's great truth may be even expanded: yes, the heart has reasons that reason knows not, but our reasons will have to include at least the recognizable needs of our hearts. We must prepare for that. Our languages will not disappear. In the beginning was not the picture but the word. And forever too.

END OF A WORLD OF BOOKS

I think that I was never—well almost never—very anxious about the sales of my books. I wrote them, they were published, and that was—well, almost—that. But I am a historian, and history does not have a language of its own. It may—it can—be read by many kinds of people; among them by some of my book-reading friends and neighbors and acquaintances; not only by professional historians, and not only by "intellectuals" (an odd category of men and women, less distinct now than they think they are). Of course I am 92 years old and many of my old friends are dead. But I think I know that now very few of my neighbors and friends buy books, whether mine or not.

My neighbors mean much to me. I have honorary doctorates and awards from universities and states and foreign countries, but one that means the most to me is the one given to me seven years ago, as the First Distinguished Citizen of Schuylkill Township, Chester County in Pennsylvania, where I had served on its Planning Commission and where I have lived now for more than sixty years.

There is nothing similar in how public opinion is manufactured and how books are bought and read. Much of public opinion and of popular sentiment (the two are not the same, but they overlap) consist of choices presented to them by politicians and their managers. Much of the choice of books is the result of their publicity presented through printed notices, newspapers, magazines, reviews. But the primacy of the printed word is now largely gone. The primacy of pictorial "information," of selected images and news has replaced it. People are not aware of what this means. It is not that they are rejecting books. It means that the diminishing "information" about books results in the diminution of their availability. Around me and, alas, around the United States bookshops are closing. Here and there they no longer exist at all.

It is in the nature of most people to adjust themselves to circumstances, rather than to adjust circumstances to themselves. That kind of adjustment is not always conscious, but there it is. In the present trade of books it has been certainly conscious among most publishers. The decline of book-reading and the difficulties of their marketing has led to the decrease of their former, even when necessarily partial, concern with the values of the books they select to publish.

A respect for books still exists, and so remains, oddly, a respect for their authors. I am a beneficiary of that. Many men and women of my remnant neighbors as well as workingmen and restaurant owners who know me, say of me: "he is a famous author." Hah! Pish! I am not famous, and I am a historian first and an author consequently but this does not matter. What may matter is that there is a respectable but somehow antique sense about this. Faintly it suggests someone like a landowner who still does his own farming or who is known for the beauties of carnations in his garden, or a painter who makes a decent living by painting not abstractions but somewhat old-fashioned canvases. (Consider that "old-fashioned" was an often negative adjective among Americans eighty years ago, though no longer now.)

But *no longer* is the appropriate phrase now. The world of books still exists, with its effects here and there; but, really, it belongs to an age of no longer. It was part and parcel of an age that began about six hundred years ago, and later called—imprecisely—the Modern Age: together with reading, with the so-called "enlightenment," with the rise of the bourgeoisie, with liberalism, with great cities, with the predominance of well-being, with money, with the retreat of barbarians, etc., etc. (The word "liberal" was praiseworthy beyond politics; in England it has been approbatory. For the last sixty years "conservative" in American politics largely means anti-liberal, with some reasons: but are the majority of American "conservatives" book-readers? I am inclined to doubt that.) But I am a historian and not a prophet. But now I must add a tiny grain of speculation to all of this. There may be even more to this devolution than "decadence." "In the beginning was the Word," not the picture. And now the end of the world of books, perhaps even of the end of reading, may have deep consequences in the relations of all kinds of people with each other: the fading of intimate mental relationships, of talk, of conversation, of *intelligence*—the meaning of a word with which I shall conclude the lamentations of this essay.

. . . .

Throughout my life I was surrounded by books. My father had an impressive library. To write about how and when and what I started to read, the atmosphere and the presence and the scent of the books, and the argle-bargle about them would be *de trop*; I have (or at least so I honestly think) not much nostalgia for old Budapest or Vienna or the Paris of Proust or smoky London or what not; I am not—well, not much—longing for The World of Yesterday; I am vexed by the world of today. Still what I had here with myself is— still—a goodly part of myself. I *am* surrounded by books, about 17,000 of them, in a splendidly high-ceilinged room in a handsome house and a garden and a family, a library that the, sometimes thoughtless, generosity of the United States of America allowed me to build, to have.

For more than sixty years I have written about thirty books, printed by reputable publishers. Their subjects—rather than their style and philosophy—differ much, which is probably the reason why I must have no concern about their potential rediscoveries, readers of them after I will be gone. But I have been concerned about my library, of what would happen to it. It contains a few valuable collections but otherwise it is an eclectic *bibliothèque*. About fifteen years ago I was told by a book dealer that no book dealer or seller could take on such a library to sell: but that there was a way out. If a university library would accept and store my library as a gift, it would have to be appraised by an outside expert, and their sum-total value could then be a substantial tax deduction for my heirs. One university with which I had been associated for many years accepted this offer. But in 2000 he canceled our agreement. He had to cancel our agreement. There is now no place for my books in their library. The world of libraries has changed. They now have to discard books, more than they acquire new ones. But then the learned director of another major university, Notre Dame, visited me and said the library would be pleased if I were to give them my books and papers.

So there with my papers too. I do not mean the manuscripts of my books but something else. I was (and still am) a correspondent. So I kept letters I received and sometimes solicited and copies of some of my letters in folders entitled "Academic and Literary Correspondence" (to distinguish them from other personal folders). For more than sixty

years, tied in rubber bands every six months or so, including some letters from a few famous men and women. Of
the very few correspondents whose letters I found excep-
tionally valuable, indeed precious, keeping them in separate
folders, was George Kennan who honored me with his
friendship, and with whom I exchanged about four hundred and twenty letters through fifty years, 1952 to 2003,
of which more than half he wrote to me, many of them
handwritten. A portion and excerpts of this correspondence
were published in the New York Review of Books, entitled
"Wise Men Against the Grain." Kennan's family told me
that this correspondence remains mine, and I could do with
it what I will.

Such have been evidences of the end of the world of my
published books, my manuscripts, my library, my letters
and correspondences. They will now be dispersed. (John
Keats, perhaps the finest English writer of letters, wrote "A
Dissertation of Letter-Writing" a little less than two hundred years ago.) As for correspondence, I will no longer
order personal stationery, since I would be obliged to have
printed on the top of it at least five items—my mailing address, my telephone number, my cell phone number, my
FAX number, and my e-mail address (which I loathe to use).
So much for the simplification of communications.

I thought of these matters without, at least so I think,
much of self-pity, but on a dark early December afternoon
I knew that I was now living not *at* but *beyond* the end of
a great Age. I had my first stiff drink and would have a second which does me some good momentarily but also tends
me to stumble, losing my balance. This is not an Old Wives'
Tale. Only the tale of an old man. At nights last week I had
re-read the "Old Wives' Tale" which is Arnold Bennett's
only very good book, not good nutrition for my mind now,

full of scenes of rain as his pages are. In accord with his somber book I was now inclined to write: "All day it rained."

But it didn't.

. . . .

I shall write no more books, not because of despair but because of my age. (I travel hardly at all, read less, search for words and names, etc., etc.) But a week after Thanksgiving an extraordinary book was published. Extraordinary: because of the riches of its contents and because of this shining style. Its title, "Inventing Wine," is somewhat deceiving, because it is more than that: a superb and concise history of wine-growing, storing, selling, tasting—a history of wine. This morning I am told that it is selling very well, which it deserves. It was written—it *is* written—by my son. Paul.

Last year I went to a funeral of a friend. Ramsay was a quiet, humble, modest man—qualities more profound than manners (Goethe was right: there are no manners which do not have something of a moral foundation). Ramsay was a veterinary doctor, a near neighbor of mine. For many years we served together in the government of our township. About twenty years ago he and his family moved away, further west, beyond suburbanization. Now my daughter drove me to his memorial service. I thought that there would not be many people beyond his family whom I wanted to see and talk with; but as we approached their house I was stunned to see hundreds of cars lined up in a field beneath it. Then there must have been at least four hundred people around and under a tent and then, after the service, in the house. It was a bright December morning. Two days before we had three inches of snow but here, less than thirty miles to the west, there was none. I kept looking out from the tent. There was a dun brown breast of a large hillside beyond, and closer what I could see of the fields and fences where he kept horses. Ramsay was one of the gentlest men I have ever known. Now all of these men and women had come to remember him. Most of them must have been his neighbors of his last twenty years. They must have come to pay respect to his qualities. What I saw was a panorama of American decency. I only knew a few of them; but I knew.

Well: "intelligence" means more than—and perhaps not at all—the ability to read. It has nothing to do with "inter-legere." Look it up in the Oxford English Dictionary. It was, and still is, the faculty of understanding.

So, to wit La Rochefoucauld: "Things are never as bad—or as good—as they seem."

THE YEAR 1924

More than once I have taken exception to how (in 1694, but then again in 1935) the Dictionary of the French Academy defined History as "the narration of actions and matters worth remembering." But: what is "worth remembering"? Everything is a potential historical source. Every human person is a historical person. Elsewhere in some of my writings I suggested another distinction: some events may have been important, others significant. What is important is actual, it had, or has, immediate consequences; what is significant is potential, later ones. Yet for this essay, these ruminations about some things and people in the year 1924, that distinction is not exactly what I have in mind. The meaning of some events in 1924 was significant, true: but I am interested, too, in their coincidences that were perhaps more (or less) than "significant."

Coincidences are spiritual puns, as Chesterton once put it. Of course coincidences are time-bound, as a matter of fact, chronological. Much of this essay is about Americans and Russians, America and Russia, in 1924. Early in that year Vladimir Lenin and Woodrow Wilson died, less than two weeks apart: Lenin on 22 January, Wilson on 3 February. (I was born nine days after Lenin's and three days before Wilson's death, on 31 January 1924.)

Both of them had been ill for some time. Lenin had a stroke in 1922, another, more severe one, a year later. I should like to know more about his funeral, but this is not a research article for which I should read about a dozen newspapers, including Russian ones, a language that I do not know. I know that it was a state funeral, even more than a Communist one (though the front pallbearer was Dzerzhinsky, chief of his secret state police). I also know that the news of Lenin's death was less sensational than Stalin's, twenty-nine years later. The main reason for this was of course his well-known illness. But there was another reason too. This was the dying down of Communist revolutionary attempts in Europe and in the Near and Middle East.

Yet Lenin's memory, his image, his legend lived on. The enormous modern mausoleum, erected on Red Square in Moscow, stands there still. For almost a century, all across the world, Communists and their sympathizers revered his name. So did some non-Communists. During the Second World War, in 1942, LIFE magazine (owned and directed by Henry Luce, a prominent Republican) printed a full-page picture of Lenin, with this caption: "This was perhaps the greatest man of the century." A few years later some intellectuals began to distinguish him from his brutal successor Stalin. Hannah Arendt, in her *The Origins of Totalitarianism*, in 1948 wrote: "At the moment of Lenin's death the roads were still open . . . Stalin won against Trotsky who...had a far greater mass appeal . . . Stalin changed the Russian one-party dictatorship into a totalitarian regime . . . It seems clear that in . . . purely political matters Lenin followed his great instincts for statesmanship rather than his Marxist convictions." The very opposite was true. Lenin's famous testament, warning the party about Stalin, was a forgery, passed on by Trotsky to an American communist, Max Eastman of New York. There is enough evidence, and by now Russian documents, to prove that Lenin was as brutal as his successor Stalin, ordering the murders not only of actual but also potential opponents of Communist-Soviet rule.

In a most important sense Lenin was a lesser man than Stalin—certainly in his statesmanship. Lenin, who lived abroad often, knew more nations and people and their histories than Stalin knew: but he understood much of that not at all. The former Caucasian bandit Stalin, who hardly ever lived for long outside Russia, eventually became a statesman with an increasing knowledge of history. Before the Bolshevik revolution in 1917 Russia, allied with France and Britain, was among the potential victors of the First World War. With Lenin and his Bolsheviks in power in St. Petersburg, Russia became a loser. After the First World War Lenin gave up large chunks of the Russian empire in Eastern Europe. (After the Second World War Stalin regained all of that, and then some.)

At the bottom of their different statesmanship was one profound difference in their view of the world. Stalin did not think much of International Communism. Lenin believed in it. The evidence is there in the many papers and letters and speeches he drafted or dictated or signed up to his death in 1924. He believed and said that Communism and Communists were advancing everywhere. The opposite was true. Just about everywhere their attempts to gain power, or even to become a dangerous revolutionary power, were defeated. Indeed this was happening even three or four years before his death, even before his first stroke. He believed, as did Trotsky, that the triumph of Communism in Russia would soon repeat itself in the countries and nations of at least Eastern and Central Europe, very much including Germany (fatherland of Marx). Instead: by 1920 in Finland, Estonia, Latvia, Lithuania, the local Communists were crushed, their movement disappeared; in Poland a national army defeated the invading Russian Red Army; already in 1919 the ephemeral (and at times ludicrous) episodes of Soviet "governments" in Munich and Budapest were kicked out and trampled down without much fighting; in Germany the last feeble attempt of Communist insurrection flickered out in 1923. Yet Lenin still believed in the coming of world revolutions—perhaps in Asia—but there they would fail too.

Half a world away his contemporary Woodrow Wilson was struck down by serious illness in the early fall of 1919. He lingered on, often bed-ridden, for another four years and four months. He wished that his prime idea of a League of Nations, established along American principles, would prevail. But the majority of American voters refused that. Still Lenin's failure was greater and deeper and more enduring than Wilson's. At the end of the twentieth century the red banner of Communism was replaced by the old Tsarist colors of Russia's flag. And while Lenin's mausoleum is still in Red Square, the remnants of the bones of Tsar Nicholas II and of his family whom Lenin had directed to be murdered in 1918 have been disinterred and buried in a Russian Orthodox church in St. Petersburg, so ordered by a new Russian government.

Woodrow Wilson was elected the President of the United States in 1912—because of a lucky concatenation of political events: the Republican challengers of this Democratic candidate for President had—temporarily—split into two camps. Shortly before his inauguration Wilson told a friend: "It would be the irony of fate if my administration should have to deal chiefly with foreign affairs." "Fate" or not: that was to be Wilson's destiny. In the short run: with disastrous consequences. In the long run: with everlasting consequences, accepted by four or five generations of Americans, prevalent even now, an entire century after Wilson became president.

Of "foreign affairs" Wilson knew little, and understood even less. In order to illustrate this would require an entire large book. He was learned, author of a book, once a university president: but his knowledge of the world was shallow (shallow, rather than like Lenin's, essentially wrong: but like a half-truth, shallowness of mind can be as dangerous as a lie). There were great yawning gaps in his knowledge. He knew some things about British history and British institutions; of the rest of European history (and geography) he knew just about nothing. Yet in 1918 destiny appointed him, at least temporarily, to something like a savior of the world. The Western allies, with American help, won the First World War. The victors, British and French and others, had different impulses, what to do with their victory. The governing ideas of Wilson (and, we must add, of many Americans) were simpler. This war was to be the end of all wars. God's decision to have American ideals prevail over most of the world was an apotheosis of American (and largely American Protestant) Progressivism. Volumes of statements of such beliefs exist in the speeches, sermons, articles, books, and other preachings of the period. Reading them now they are astonishing because of the shallowness of their pompous naiveté (naiveté, rather than mere innocence), resounding through their megaphonic phraseology. They are worse than laughable; they are shocking. And so are and were many, perhaps most, of Wilson's own statements. Immediately after the end of the war he went to Paris twice, staying there for long weeks, even months. His main idea and obsession, that of a League of Nations, constructed with American ideas and with an American presence as its main strength, had Republican critics at home. He would not compromise with them; instead he spoke nonsense after nonsense about the peacemaking after the war, including the Treaty of Versailles. This treaty, he proclaimed in September 1919, "is a ninety-nine percent insurance against war." And: "at last, the world knows America as the savior of the world." "I am glad for one to have lived to see this day." He collapsed ten days later.

He had been responsible—sometimes entirely, sometimes partly—for many of the enduring blunders and errors of the remaking of Europe in 1918 and after. This is not the place to recount them. Instead, let me ruminate, if only for a moment, about "the irony of fate." Had it not been for the First World War and "foreign affairs" Wilson would have been a mediocre domestic president, causing not much harm. Like Lenin, the sources of his ideas were often those of the century then past, many of them British liberal (though without the depth of his part-time idol Gladstone). But there is another Great Irony of Fate, even greater than— whether fate or destiny—propelling him into the First World War. This is that Wilson's ideas and ideals lived on and on, adopted by many, perhaps even the majority of American people, after his death to this day. A century ago his most intelligent opponent, no matter what his faults, was Theodore Roosevelt. In 1912 the petty politics of the Republican party deprived Roosevelt from being their candidate, the prime opponent of Woodrow Wilson, whom he would have defeated, as the three-way electoral results of 1912 prove. Roosevelt had his shortcomings, but he knew and understood some things in the world definitely better than Wilson. Had Roosevelt been elected in 1912 and thus been president in 1914, the entire history of the First World War would have been different. In that event he would have bequeathed to the American people a conduct and a philosophy of foreign relations other than Wilson's. But that was not to be. Wilson's views of the world eventually became the American view of the world, adopted by such different people as Herbert Hoover, Franklin Roosevelt, John Foster Dulles, Richard Nixon, Ronald Reagan, Clinton and George W. Bush. Franklin Roosevelt knew more of Europe and the world than Woodrow Wilson; yet he thought, too, that the United States should have been in the League of Nations as Wilson had wanted it, and, if so, there would have been no Second World War—at the end of which FDR's principal aim was to establish a United Nations, a greater and improved version of Wilson's League, to ensure world peace, which of course it did not. When Richard Nixon became president he ordered that Wilson's desk be brought up and installed in his office. In 2001 George W. Bush declared: "America must fight the enemies of Progress"—a view of the world and slogan that could have been uttered, word for word, by Wilson nearly one century before.

Why do men fail? Almost always because of the faults, or shortcomings, of their thinking. That has little to do with Intelligence Quotients, brain conditions or whatnot. Most people do not *have* ideas, they *choose* them. Lenin was a Russian intellectual. He too believed in Progress—which, to him, was only retarded because of the existence of classes: that classes meant more, they were more of a reality than were states, or than nations. The inequality of classes must be crushed by Communism. In sum, Lenin believed that history was made, that it was the result of material economic conditions: a nineteenth-century idea of human nature that was, and remains, essentially wrong. He was outdated; and so was his contemporary Wilson. But then a man's choice of his ideas has something to do with his character. The flawed essences of Wilson's ideas were hardly independent from his character, flaws probably deep-seated as well as shallow. Wilson was often petty and even vindictive. But he was not brutal like Lenin. He was not Russian but American. Outdated they were both.

Succeeded they were by Mussolini and Hitler, outdated not at all. The former was already in power in 1924, the latter in 1933, a man who another seven years later came close to winning an entire Second World War. Neither of them believed in Economic Man. They thought in terms not of the struggles of classes but of nations. "The violent bear it away."

. . . .

We are not finished with 1924. It is only February, and less than ten days after Wilson's passing away on 12 February there was a significant happening in New York City. This was the performance of George Gershwin's "Rhapsody in Blue." It came at the end of a much advertised ambitious program of New American Music.

It was an instant success. New and startling as it was, the ears and the minds and the hearts of the audience were ready for it. They poured out under the brilliant lights of Aeolian Hall into the winter streets of New York, lit up too, toward and into their big-wheeled automobiles, in what they sensed and thought was the center of the world, the modern world. They were not altogether wrong. 1924 in New York was the peak of the Jazz Age. It was also the starburst triumph of American music for much of the rest of the world, an audible and palpable symbol of what would be the American Twentieth Century.

Its composer George Gershwin was a musical genius. He wrote "Rhapsody in Blue" in a few days, requested by Paul Whiteman whose massive orchestra performed it. The reactions to "Rhapsody in Blue" of the music critics of the big serious newspapers were mixed; that did not matter. A year later Gershwin wrote his more substantial "Concerto in F," and then many other things. But where his genius came, what crystallized and formed it, is a part of a wider and deeper story Jews, Russians Jews, in America, part and parcel of the history of Russia and America and within the latter, a stunning appearance of Americanization. Lenin and Wilson were dead; but only a few days after their funeral a young Russian Jew appeared to represent America to itself, and to the world.

Russia and America: ninety years earlier in a short paragraph Tocqueville wrote that the Russian and the American empires would be the greatest empires to come. They incarnated wholly different political and social systems, but their relations were and remained good for a long time. After about 1880 there came a change, but not on the strategic or diplomatic or state level. Eighty years earlier most of the Jews of the world lived within the confines of the then Russian empire. During the nineteenth century hundreds of thousands of them moved westward, mostly to the cities and towns of eastern and central Europe. After 1880 more and more of them were able to transport themselves to America. The Open Door practices of the United States as well as the indifference of the American people allowed them in. Soon there were new developments. More and more Americans became uneasy at the sight of these strange aliens crowding into their cities. Other Americans were aware of the political backwardness of the Russian Tsarist regime. When in March 1917 a revolution put an end to Tsarist rule, Wilson said that this was one of the greatest events in world history. It contributed to his decision to enter the European War which he now thought was a perfect alliance of Democracies against Imperial Germany. As almost always when it came to foreign affairs he was wrong: indeed soon he became a main proponent of a decision to intervene in Russia and help crush the Communist revolutionary regime. All of this mattered not much in the long run. American-Russian relations continued, even though officially the United States did not give diplomatic recognition to Soviet Russia. Yet in the midst of the American Red Scare, President Harding said in 1923 that "there is an unfailing friendship in the United States for the people of Russia."

What mattered were the new American immigration laws of 1921 and 1924, restricting severely the immigration of more Russians to the United States. But by that time the presence of Americans of Russian Jewish origins was considerable politically too. The British Balfour Declaration in 1917, promising a homeland in Palestine for Jews, was made, among other things, with American Jews in mind. The Communist Party of the United States, formed soon after the Bolshevik Revolution, was mostly composed by Russian Jews (some of its early meetings were conducted in Yiddish).

But the children, the sons of these Russian Jewish immigrants, had nothing to do with Russia. They and their language was Americanized extraordinarily fast. That was the development of many different immigrants' children too; but seldom with the dynamic speed of those of Russian Jewish origins. Startling evidence of the acculturation—or more precisely: Americanization—of some of them was their appetite for and their contribution to their composition and production of American popular music. To write about music is not (or, rather, should not be) easy. Words may have their own music, but their meaning is not music. Poetry can exist without reading it, but music without hearing it cannot. What interests me is something of which the results are obvious, while their sources are mysterious.

How is it that in the heads and hearts of these Russian Jewish youngsters there arose sounds and their component embroiderings and then songs that were, and became, quintessentially American? Nothing arises from nothing and, yes, there were Negro (but also Anglo-Saxon) elements in American popular music: but the sounds, tones, tunes that Gershwin and Arlen and Berlin and Kern (for once an Austrian, not a Russian Jew) wrote were something new and something else. Their beauty existed less because of their melodies than because of their astonishingly sophisticated chords, harmonies. In these harmonies there was often a melancholy beauty: but there was (contrary to some of their heavy-handed exegetes) nothing Jewish, nothing Russian in them. Perhaps inherent in them were their composers' aspirations to an Anglo-Saxon, and largely feminine elegance of an upper class that was reachable in America, unlike in Russia. And the lyrics superimposed on those songs only contributed to their success which was soon overwhelming and world-wide.

The reign of this kind of American popular music came to an end around 1950. What went on was the presence and influence of Jews in American life and history. Another coincidental date: by 1950, for the first time in history, more Jews lived in the Western Hemisphere than in Europe. In the 1960s and 1970s there was an abundance of Jewish writers within American literature but that too did not last. Until well after the Second World War most Jews in America were pro-Russian, but then the so-called "neoconservative" wave emerged, including even more American Jews of Russian fathers and grandfathers: anti-Communist and anti-Russian they now were. Their influence was more than considerable but that faded too. What did not fade was the American political (and sometimes and in some places popular) American commitment and alliance and support for the state of Israel (an inclination some of the roots of which reached far back into American religious and spiritual history). The history of the Jewish element and factor in American history is a long and complex one; and, within it, so is the influence of Jews in the long history of American-Russian relations.

....

1924 was the end of the immediate consequences of the First World War. Turmoil and armed struggles and large and forced population displacements and revolutionary attempts went on for five years after 1918. In Germany the last revolutionary attempts and an enormous unprecedented inflation ended in November 1923. If we can speak of a relatively peaceful intermission between the two world wars, these were the years from 1924 to 1933. Yet that intermission involved international politics but little else. The First World War was followed by immediate changes in mores, manners, fashions, technics, arts, architecture, literature, pictures, entertainment, music wider and greater than anything before. They came about almost everywhere in the world, in the winner as well as loser nations of the war. In the United States their effects were soon called The Jazz Age. That, as we shall see, is an imperfect title and perhaps even an imprecise definition: but there it was, and we may designate its peak in 1924 in the United States. What happened on that evening of 12 February 1924 in Aeolian Hall in New York City was a significant symptom, perhaps even a mark of it. But it was a significant event within a larger scene. By 1924 New York was the Modern City of the World, and the United States was the most Modern Country of the World. That word "modern" was used and misused earlier too, for all kinds of purposes and in all kinds of ways, but by 1924 its connotation had become large and definite. "Modern" had become a universal approbatory adjective in the American language, though not quite yet in English in England. "Old-fashioned" was a pejorative adjective in America, though not quite elsewhere. Of course the relations and the behavior of men and women changed too, but that was a longer and deeper development, useless to pinpoint it to 1924. Yet while to pinpoint the peak of the Jazz Age to 1924 and to New York may not be precise, an erroneous exaggeration it is not.

But the United States was, and is, a vast country; and there are dualities—sometimes even a recognizable split-mindedness—in the thinking of Americans too. Enormous portions of the United States, large majorities of its people were unaffected, untouched by the Jazz Age. Much of American life and civilization had become urban and metropolitan, no longer rural or even small-town, industrial and no longer agricultural: but it was not along these categories that American minds were divided. In 1924 New York City and Sauk City were still very much apart. The music and the letters and the fashions and the mores of the Jazz Age penetrated the Midwest, the West and the South too: but frequently in dribs and drabs. Automobiles, telephones, radio and the movies: yes, penetrating the vast country gradually. Certain entertainments: yes, changed; there were changes in clothing; but the mind-set of a great majority, their modes of thinking and their professed beliefs hardly at all. This is why the categorical definition of the Jazz Age is imprecise. The very minds and the preferences of the American people were divided, not for the first or the last time in their history. An obvious example of this was their political elections. Surely Calvin Coolidge was not a denizen or a participant of the Jazz Age; but he *was* one protagonist of the Twenties. In 1924 twice as many Americans chose him for their president than they chose his opponents; he also carried New York City. Just as telling was, and is, that the Democratic candidate (John W. Davis) may have been his competitor but not truly an opponent: his platform, whether domestic or foreign, whether financial or populist, was not very different. Significant, too, was the diminished popularity of the Progressive party. Their impressive candidate for a Third Party, Farmer Labor, Populist-Progressive Robert La Follette, got less than one-third of Coolidge's votes even in the Midwest, and carried only one state, Wisconsin, narrowly. In 1924 and indeed throughout the Twenties political radicalism was unpopular and un-American, save for a new kind of racialist nationalism, having moved from the South to the north and westward. The attraction and the membership of the Ku Klux Klan reached its peak in 1924, evident even in such states as Pennsylvania or Ohio. Many American working men who had been the hope and sometimes the mainstay of Socialist and Progressives a decade before, in 1924 supported the Ku Klux Klan. That did not last long but it was a symptom—as was the dominant popularity of Calvin Coolidge, who in 1924 proclaimed that the issue was "whether America will allow itself to be degraded into a communistic or socialistic state or whether it will remain American." That was not an issue in 1924 or ever since: but anti-Communism as the main ingredient, indeed, as the essence of American patriotism had become a principal force propelling candidates into presidencies, including Ronald Reagan, who sixty years after Coolidge chose to display Coolidge's portrait in the presidential office in the White House.

Yet instead of Coolidge's, Reagan should have displayed a portrait of Louis B. Mayer (another super-American and "conservative" Jewish movie magnate) on the walls of the White House. Reagan owed everything not to Calvin Coolidge but to the movie industry. Here we turn, for the last time, to the split-mindedness of the American Twenties. Progressivism and its erstwhile political and intellectual proponents had shot their bolt by 1924. But what was (and remained) unquestioned and unquestionable, powerful and dominant, was the American idea and belief in Progress, a belief as current in Sauk City as in New York City, unquestioned by George Babbitt as well as by Sinclair Lewis. A few thoughtful Europeans and Englishmen (Douglas Jerrold, Hermann Keyserling, Johan Huizinga) visiting the United States in 1926 or 1927, were startled and appalled by the unlimited popularity of the American idea of Progress. That was not a twentieth but an eighteenth-century belief and dogma, they wrote. It was both juvenile and senile, exemplified, for instance, by the wizened Henry Ford. ("History is bunk." "Books muddle me.") In 1928 Julius A. Klein, Herbert Hoover's Assistant Secretary of Commerce, declared: "Tradition is the enemy of Progress." This man was not a liberal or a Democrat, he was an American conservative (as was George W. Bush more than seventy years later: "America must fight the enemies of Progress.") Coolidge in January 1924: "After all, the chief business of the American people is business." The dean of the Divinity School of the University of Chicago: "Business is the maker of morals. What else but business can make morality?" A most popular nonfiction book in 1925, Bruce Barton's *The Man Nobody Knows:* "Jesus Christ was a virile go-getting he-man of business, the first great advertiser, a premier group organizer, master executive, a champion publicity-grabber." The president of the Federal Council of Churches of Christ in 1924: "Moses was one of the greatest salesmen and real estate promoters who ever lived."

The subject of this essay is the year 1924, not the Twenties: still I am compelled to say something about the qualities of that American decade. It may be argued that the Twenties were the *only* Modern Decade of the twentieth century: it was then that "modern" literature, music, painting, architecture, fashions, habits spread rapidly worldwide. (The fake "revolutionary" 1960s were nothing more than an artificial, and largely juvenile recapitulation of the 1920s.) And the American "isolationism"— the "back to normalcy" (Warren Harding's phrase) of the 1920s—was far from complete. Americans rejected the League of Nations, and a president such as Coolidge ("a New England backwoodsman," Winston Churchill then said) was instinctively hostile to and wholly uninterested in Europe; yet in the 1920s American art and artists were more involved in Europe than ever before. In 1924 European immigration to the United States was largely curtailed; yet more actors and performers from Europe arrived in the United States than ever before; and across the Atlantic were shipped masses of European objects of art, including sometimes entire buildings. "Conservative" was an adjective avoided by most Americans and especially by their politicians in the Twenties, and at least for two decades thereafter; but after 1950 that designation became popular and at least some of the inclinations of American "conservatism" of the 1920s have continued to prevail to this day.

In January 1924 Great Britain, for the first time in its history, had a Labor government and Labor Prime Minister. This meant that the century-long domination between the British Conservative and the British Liberal parties (and their preceding competition between Tories and Whigs) came to an end. Labor, a third party and a third force, kept creeping up the ladder of British elections; after the war more and more Liberal voters chose Labor. In 1922 the Liberal war leader and lion Lloyd George left the political arena even before an election during which now more people voted Labor than Liberal. In December 1923 there had to be another national election in which the Conservatives still had most of the votes, but with few Liberals joining them (Winston Churchill did) the Conservatives did not have enough for a majority. Now the Labor votes and parliamentary seats well outnumbered the Liberals'. Now they had become the second largest party in Britain. The judicious King, George V, agreed that Labor must now have "a fair chance." (He also wrote in his diary that night: "Today 23 years ago dear grandmamma [Victoria] died. I wonder what she would have thought of a Labor Government.") The Labor leader Ramsay MacDonald became the Prime Minister of what was then still the greatest empire in the world—on the day Lenin died, 22 January 1924.

Whether that coincidence was a "spiritual pun" or not, it was a milestone in British history. A milestone, yes: but a turning-point it was not. MacDonald's government was cautious, moderate, slow, contrary to the fears of many conservatives as well as to the expectations of most radicals. The Labor regime was not able to do much, if anything, about unemployment which was considerable in Britain in 1924. It was also pelted by attacks against its relatively moderate politics in regard to the Soviet Union. Before the end of the year the Conservatives swept back into power, with the Liberals crumbling even faster than before. They had become a minor third party ever since. Once more during the interwar years MacDonald and Labor were permitted to govern but again they soon failed. It was absurd to even think that one day Labor would be the steadiest supporters of a man such as Winston Churchill. Yet honor and glory to them, that happened on 10 May 1940 when Churchill could not have become the British Prime Minister without Labor's unstinting support. By that time the League of Nations had ceased to exist, and Lenin's corpse was moldering in his mausoleum.

In 1924 few people outside of Germany and Austria heard or knew anything about Adolf Hitler. For most of that year, about eleven months, he was in a prison in Bavaria. Yet that year was one of the turning-points of his career. What happened with him, and within him, turned out to be significant. On 9 November 1923 he and his National Socialists and other extreme proponents of German nationalism had attempted an uprising in Munich, a "National Revolution," hoping that it would spread across much of Germany. The uprising failed, because—after some hesitation and tergiversation—a few leaders of the army and of the Bavarian government resisted it. There was a short burst of gunfire. A few of Hitler's friends were shot and killed, others were wounded, most of them escaped. Hitler survived, suffering from a broken shoulder. He fled but then he was arrested. On 24 February 1924 he was brought to trial before a state court. He turned the trial into a rhetorical victory for himself. He defended nothing, admitted nothing, denied nothing: indeed he proclaimed the cause of the National Revolution: "Yes! We wanted to destroy this state!" "I want what is best for the *Volk*!" The judges were uneasy. The courtroom was crowded with sympathizers. The three members of the lay jury, on whom the sentencing depended, were inclined toward Hitler. In the end they agreed not to pronounce "not guilty" only because the presiding judge told them that Hitler's sentence would not be severe, five years in prison but with Hitler's opportunity for a parole within a year.

Adolf Hitler was not—or: surely very, very seldom—a happy man. Yet 1924 may have been the happiest year of his life. The conditions of his "imprisonment" could not have been better. He was lodged in a former castle, Landsberg, in Bavaria; the director of the prison was sympathetic to him; he could move around, receive visitors, dictate to a secretary; receive letters and bouquets of flowers every day. His warders admired him. His behavior contributed to the prison director's recommendation to grant him parole, and soon. But the most fortunate circumstance, nay, development of his prison months was his leisure, his self-employment of much reading and some writing. He read a great lot, all kinds of books brought or lent him by friends. Landsberg, he said, was his university. "The state paid for my higher education." And it was then and there that he wrote—or, more accurately, dictated—his soon famous "Mein Kampf." He was a speaker, not a writer; he was very conscious of the differences between the spoken and the written word (he once said "Mein Kampf" ought to be spoken rather than read). That book is remarkable perhaps less because of its often cited extreme ideological and political propositions than because of his first autobiographical portion (in which Hitler twisted and obscured some matters) but which is a clue to his personality and character. In 1924 the political prospects of his National Socialist party or movement were not too good. The five years of postwar political and financial crises in Germany seemed to be over by the end of 1923. Thirteen days before Hitler's release from prison the small extremist *völkisch* party won only 3% in a German national election. But Hitler was not dismayed. He said that he had been the Drummer of a movement; now he would be its Leader. He now realized that revolutions or uprisings were not the ways to gain support among Germans. He would rise to power democratically, as leader of a Party of National Order. That came about a little more than eight years later.

And here and now I must draw my readers' attention to something which involved (and still involves) more than 1924 and more than Germany. In Britain at the end of October 1924 there was a new election. Mostly because of the decrease of the Liberals, the Conservatives now had enough of a majority to replace MacDonald's Labor government. On 9 December the King opened a new British Parliament. Labor had now replaced the Liberals but the British two-party system remained solid and unshaken. The great and long dialogue and debate and competition of the nineteenth century between a moderate Right and a moderate Left in Britain continued to prevail; it prevails even now. But—contrary to Hegel's theory of thesis-antithesis-synthesis—in many places, and not only in Germany, Conservatism and Liberalism had begun to be succeeded by two new great forces, Nationalism and Socialism. Hitler knew that very well. The very name of his National Socialist party combined the two. Observers and opponents called them "Nazi-Sozi." Socialism (and capitalism, too) he said must not be international but national. Already in 1924 the "Sozi" part of the designation or adjective faded and disappeared. "Nazi" remained, and remains with us still.[1]

---

1   Consider that this coexistence, or combination, of nationalism and socialism marks governments and parties in much of the world even now. Hitler's German National Socialism was one, extreme, variant of this.

Hitler left his "prison" on 10 December 1924. Thereafter his rise to power in Germany began. In 1924 it seemed that, for different reasons, both the United States and the Soviet Union had withdrawn from Europe. But fifteen years later Europe became the center of world history again. In 1939 Hitler's Germany started the Second World War. In the end it took the combined might of America and Russia (and Britain) to crush it.

HITLER AND "AMERIKA"

During what is now two-thirds of an entire century after Adolf Hitler died there has been no significant attempt of a historical "revisionism" or even a partial rehabilitation of this character and record. The only marginal exception to this are the works of David Irving whom people dismiss because of his obvious political and ideological inclinations, even without a careful examination of Irving's "documentation" where they could find ample evidences of his twisting and even falsifying documentary sources. At the same time there are reasons and evidence to question or at least to reexamine the often simplistic categories applied to Hitler. He was not a simple person. That ought to be obvious. And evident should be, too, the many examples suggesting the ambiguities of his vision and of his opinions. Some of the most telling, and at times startling, examples of these may be found in his changing and often contradictory views of his opponents, very much including foreign nations and their statesmen.

Before as well as after his assumption of power in Germany he believed and said that his Germany should never fight Britain; that the British may understand that he wants their Empire untouched, while they should accept his domination of much of Europe (or at least of Central and Eastern Europe). By early 1939 he realized that this would not work. Thereafter his "love/hate" inclinations toward Britain (a primitive phrase, but let it go) changed. The "love" disappeared and often not only hate but contempt for Britain and the British took place in his mind. His conviction that Britain was now his principal enemy went on, even explaining his decision to invade Soviet Russia in June 1941. There are ample evidences of this,[1] both before and during his invasion of Russia. Many people, including even historians, believe that that was another foolish mistake of an obsessed fanatic whose motives and purposes were to conquer "Lebensraum" for the Germans while eliminating the inferior Slavs. That monocausal explanation is insufficient. In 1941 his main plan and hope was that after conquering most of European Russia and defeating Communism, Churchill and Roosevelt would be forced to recognize his invincibility, and the British and the American

---

1 To General Franz Halder, *Kriegstagebuch*, War Diary, 14 June: "After the attack on Russia and the evolution of his calculation that the collapse of Russia will induce England to give up the struggle." On the same day: "The main enemy is still Britain." (From an unpublished source, a Luftwaffe general, in Irving, "Hitler's War," p.266.) Walter Hewel's Diary (Hewel was not a general but very close to Hitler, killed himself after Hitler's death): 19 May 1941: "When Russia is defeated this will force England to make peace. Hope this year." Hewel's diary, 20 June 1941: "A long conversation with the Führer. Expects a lot from the Russian campaign. He thinks that Britain will have to give in." To Field-Marshal Keitel, his chief of staff, 18 August 1941: (ADAP: Collection of German foreign policy documents D, 13/1, p.346): "The ultimate objective of the Reich is the defeat of Great Britain." Halder's War Diary: Kiev, closer and closer to Moscow: Hitler's aim is "to finally eliminate Russia as England's allied power on the continent and thereby deprive England of any hope of a change in her fortunes." To Admiral Fricke (War Diary of the Navy) 28 October 1941: "The fall of Moscow might even force England to make peace at once."(See also my *The Hitler of History*,1997, pp.150–51.)

people would have to consider some kind of an accommodation with him. That was neither foolish nor nonsensical.

But sometime in mid-November 1941 there came a change, a change in his entire vision of the war, and of his own destiny—seldom adequately noticed or emphasized by historians. He realized that he could no longer win this war on his terms. Evidence of this is there in General Halder's War Diary on 18 November 1941: "The recognition, by both of the opposing coalitions, that they cannot annihilate each other, leads to a negotiated peace." And: "We must consider the possibility that it will not be possible for either of the principal opponents to annihilate the other, or to subdue them entirely." Note that he said this before the Russians began to push the German army away from Moscow, as well as before Pearl Harbor. Now he thought—and directed—Germany's preparation for a long war. The "Blitzkrieg" period of short and decisive offensive campaigns was over. But he did not say, or believe, that his war was lost. His idol was now Frederick the Great who, facing an unprecedented and overwhelming coalition in 1756, was convinced that their unlikely alliance against Prussia must be broken up, mostly by defeating one of their armies, which is what happened. And that too was Hitler's entire strategy until the end of the war.

This brings us to the United States, which now began to preoccupy more and more of Hitler's mind. His vision of the United States was ambiguous. And this is a large and important theme through which the dualities of his opinions and inclinations are frequent and telling, both in the short and in the long run.

There was nothing very extraordinary in Hitler's views and decisions about the United States before 1939. The ambiguous nature of his opinions prevailed here too. He read much about the United States in his youth, especially and almost all of the books written by Karl May whose American adventure stories were a frequent diet for an entire generation of Germans and other Central Europeans. There was nothing anti-Semitic or racialist or anti-American in May's writing. May (who never visited the United States and whose knowledge of English was fragmentary) died in 1912 when Hitler was 23. He took himself to May's funeral, but that was not the end: he kept re-reading him for many years. (It may be interesting that during the Second World War, in October 1944 he and Heinrich Himmler ordered a reprinting of 300,000 copies of Karl May's books, to be distributed among Germany's fighting troops.) During the First World War Hitler was often up at the front against British and French soldiers, but hardly ever against American ones. Not until after he had begun his political career in 1919 was he told by some of his supporters to consider the United States seriously. Some of these were Germans (for example the journalist Colin Ross who would kill himself a month after Hitler died) but also famous and Germanophile writers and explorers , such as the Norwegian Knut Hamsun and the Swedish Sven Hedin, who knew America and later visited Hitler. There was his ambiguity: his respect and even admiration for American industrial organization and production (whence, too, his praise for Henry Ford in the 1920s and for Charles Lindbergh in the 1930s); yet at the same time he was contemptuous about the changing character of the American population. For a long time he believed that the working habits of Americans were largely due to the predominantly "Nordic" character of their people. At times he was skeptical of the entire worth of American mass culture.

Such ambiguous judgments were held by many Europeans in the 1920s. But then in 1933 Hitler rose to become the master of Germany—almost at the very time when Roosevelt became the president of the United States. Of course Roosevelt had no sympathy for Hitler; he also disliked Germany and Germans, an inclination that began early in his life when he visited Germany in his boyhood. But there was no particular crisis in the German-American or even in the Hitler-Roosevelt relationship until early 1939. Hitler and Goebbels ordered the German press not to be unduly critical of the United States. Their support of German-American organizations within the United States was marginal and limited. The report that during the 1936 Olympics in Berlin Hitler refused to shake hands with Jesse Owens, the winning American black sprinter, was (and, alas, often still remains) a canard. That Roosevelt and his government and much of the American press were critical of the Third Reich was obvious; but Hitler attributed this to the influence of Jews.

Then, in January 1939, this changed. Hitler now realized that Roosevelt was his enemy. There were one or two diplomatic documents, secured by German intelligence, that were proofs of this, but there was more involved than that. Hitler now knew that Roosevelt, often secretly, began to encourage British and French politicians who were willing to fight Hitler, even at the cost of a European war. In an important speech in April 1939 Hitler was dismissive and contemptuous of Roosevelt. He would still on occasion receive and talk with American political or industrial leaders who, he knew, were no partisans of Roosevelt (Herbert Hoover, for example)—but he was not surprised by the increasingly close relationship between Roosevelt and Churchill.

Even during the stunning triumphs of the German armies across Western Europe in 1940 (which he often directed from this personal train, code-named "Amerika") he gave an unprecedented interview to an American journalist (Karl von Wiegand) of isolationist inclinations. It summed up his vision: a Monroe Doctrine for Europe, like the Monroe Doctrine for America. His Third Reich had no interest in the Western Hemisphere, while the United States should accept and understand a new "Europe for Europeans," of course now under Germany's leadership. It was his message to American isolationists; but it did not resound very far. Meanwhile Roosevelt gave more and more support to Churchill, and Hitler began to think of eliminating Russia. Roosevelt saw the Atlantic as the United States' main concern. In 1941 American Marines were sent to Greenland and Iceland while the American Navy was ordered to confront German naval craft as if these were enemy forces. Then twenty-four hours before his invasion of Russia, Hitler gave a peremptory order to his naval forces: they must avoid any combat with American warships in the Atlantic, no matter what the circumstances, even if the provocation of a fight came from the American side.

This was one of the key decisions of the Second World War. Hitler now embarked on his conquest of Russia, "England's last hope"; but his concern now was how to keep the United States out of the war. He well knew that Roosevelt would welcome a naval fight in the Atlantic, furnishing him with an argument and a pretext to enter the war against Germany; but, despite various clashes between American and German craft in the Atlantic, this did not come about. Instead, there came Pearl Harbor; the United States entered the war; and Hitler was forced to deal with this. For those who still argue that Hitler's declaration of war against the United States was but another example of his maniacal world policy, we ought to recognize that, foremost because of his alliance with Japan, he hardly had another choice.

Note, too, that he lifted his peremptory order to his naval forces not to attack or even respond to American provocations only three days before his declaration of war on the United States.

That came, as we saw, more than three weeks after his words to some generals to the effect that a total German victory in this war was no longer possible; that therefore his Germany must prepare for a long war, with the determination and hope that this Unnatural Coalition of Britain, Russia and the United States, of Capitalists and Communists, would break apart. And he recognized, too, that the United States had become the main element in that unnatural alliance against him. From 1942 to the end of the war several of his associates argued, and on occasion tried, to establish contacts with one or the other of Germany's large enemies. As Klaus Fischer judiciously put it, Hitler himself "did not initiate peace feelers, but neither did he discourage subordinates from doing so."

There was another, not unreasonable, argument he brought up against those who thought the war against one or another enemies must be stopped. Before anything, the condition of any negotiation was an impressive German military victory in the field. This was what Frederick the Great had done; this was what he had to achieve now. This was why he ordered the tremendous German armored offensive in the Kursk salient, six months after the German defeat at Stalingrad; this was, too, the reason for his last attempt of a stunning German offensive in Belgium in December 1944, the first aimed to impress the Russians, the second to stun and impress the Americans. Both offensives ground to a halt; but Hitler's wish and hope to break up the coalition of Britain, Russia and America went on—and now with the principal object of America.

About America, the ambiguity (or duality) of his own judgments persisted. Less than two months after his declaration of war against the United States he spoke positively about American industrial capacity and organization to some of his subordinates. In his now less and less frequent national broadcasts he went on speaking contemptuously of the entire civilization of the United States, but that kind of propaganda was meant for the German people. Three years later, in February 1945, close to the end of the war and his demise, he said to another small group: "The war against America is a tragedy." That was more than yet another example of his ambiguity, more than a matter of words: it suggested his real beliefs and sentiments. Beginning in early 1944 he did not discourage attempts to achieve one kind of contact or another with Americans— always also with the purpose to use the very fact of such contacts to bring about suspicions and problems between Moscow and Washington. Throughout 1944 and then in 1945 his faithful minion Heinrich Himmler (by then the head of German security) attempted to get into contact with Americans on various levels and occasions. They included, more often than not, desperate Jewish organizations in Hungary and in Turkey. They were scotched by the British but Himmler went on and on. In August 1944 he ordered the Gestapo to stop its efforts to promote the deportation of the remaining 200,000 Jews from Budapest. (About 400,000 Hungarian Jews had been already deported, most of them to Auschwitz.) In October Himmler ordered the gassings in Auschwitz to stop; in January 1945 he forbade the destruction of the last principal ghetto during the siege of Budapest. Even the record of the legendary brave Raoul Wallenberg must be understood from this angle. Wallenberg was not an official Swedish diplomat, and on his journey to Budapest, he was treated by Himmler's people and the Gestapo as a privileged (Swedish and American) envoy. Of course one of the Germans' purposes was to let the Russians know this, which is what happened. The tragic end was that after the liberation of the Pest side of the Hungarian capital (and after his having saved the lives of at least 25,000 Jews) Russians arrested Wallenberg and carried him to a prison in Moscow where he died two years later.

There is every reason to believe that Hitler knew about these attempts of Himmler, even though he may not have ordered or directed them himself. At the same time there is at least one episode of Hitler's own attempt to establish some contact with Americans. Shortly before the German retreat from Rome in June 1944, Hitler's propaganda emphasized the German respect for cultural assets such as the Eternal City. He suggested to Field-Marshal Kesselring to try to contact American generals in order to impress the world with German-American cooperation to protect Rome. But this did not come about.

The most telling and fairly documented episode of German-American contacts were the negotiations between the SS General Karl Wolff and American officials in Switzerland (including Allen Dulles), to promote an armistice between the German and American armies in Italy and the eventual withdrawal of the former from the Italian peninsula. These negotiations went on through most of 1944 into April 1945, with Wolff going to Switzerland on several occasions. They were approved by Hitler. It is telling that he had Wolff come to his headquarters in Berlin on April 16 and 17 in 1945, less than two weeks before his suicide and death. He agreed with what Wolff was doing, and encouraged him to go on, though with caution: go slow, he said, because in about two months the Unnatural Coalition between Americans and Russians was bound to break up. Did he still believe that? Yes and no. On April 13 the sudden news of Franklin Roosevelt's death arrived, and an excited Goebbels brought champagne to the Führer's headquarters, exclaiming that the turning-point of the war had now come, just as when Frederick the Great learned of the sudden death of his arch-enemy the Tsarina Elisabeth of Russia. But Hitler (who disliked champagne anyhow) did not join in the merriment. But he still thought that sooner or later the Unnatural Coalition would break up. And so it did; but too late for him.

In one important sense there was an accord between Hitler's wishes and those of the German people. They, too—with every reason—believed and hoped that they had less to fear from the Americans than from Russians (and even from the British), with consequences that a suspicious Stalin understood very well. Knowing about the Wolff meetings with Americans in Switzerland, he sent an angry note of protest to Roosevelt, whom he otherwise admired. (He was crafty enough to blame the British for this matter, which was not the case.) Another eight days later he protested again to Roosevelt about how easily the American armies were advancing through Germany while the Germans fought bitterly against the Russians across the entire eastern front. This telegram arrived on 12 April. Roosevelt chose to pooh-pooh it, but a few days later he was dead.

"Hitler and America" suggests one or two important matters. One is that, contrary to still accepted ideas (and to Roosevelt's often repeated statements), Hitler's aims did not include his eventual domination of the world; his domination of Europe and of European Russia were enough. The other is the necessary dismissal of a still often popular idea, that Hitler was insane. If this had been so, how and why can we hold him responsible for his awful acts and orders and words? Here again America's historic mission enters. In July 1941 Stalin told Harry Hopkins, Roosevelt's special envoy to him, that the United States must enter the war, because Hitler's Germany was too powerful for the Russian and British empires together to defeat it. He was right. Such was the extent of Hitler's armed might. In 1940 there was Churchill's historic mission and achievement: he could not win the Second World War but he was the one who did not lose it. Five years later the United States won the Second World War across the Atlantic and the Pacific and became the greatest power in the world. Soon the conflicts between the United States and the Soviet Union began. Hitler foresaw some of this, but about the conditions and the circumstances and the eventual development of the "cold war," about the consequences of America's primacy in the world, his hopes were wrong. There would be no Russian-American war.

AMERICA AND RUSSIA:
WHAT A STRANGE HISTORY OF OUR WORLD

"Roosevelt and Stalin" is the title of a book, recently (April 2015) published in the United States. I have not read it, but its title tells what it is about. From 1941 to 1945 the alliance of Roosevelt's United States and Stalin's Soviet Union led to the victorious end of the Second World War. Roosevelt died even before the war ended. But 1941–45 was only one chapter in the history of American-Russian relations. They changed after that. In the long run the developments of these relations ruled much of the history of the world. One hundred years ago, in 1915, the First World War was going on. During it some people recognized the importance of Russia and of the United States. Yet even then no one thought that their relations would eventually dominate much of the war. One hundred years earlier, in 1815, another great world war had just ended. Not one of the then great statesmen thought that the future of Europe would depend on the relations of America and Russia. (How could they? When, later in that century, Tocqueville, the greatest analyst and visionary of democratic America, saw much of the future of the United States and, here and there, even of Russia, he wrote nothing about their potential relations.) Yes, the twentieth century was, by and large, the American Century, on many levels affecting the lives of peoples across the world. But when it comes to traditional world history, meaning the relations of the great powers, the relations of America and Russia would soon become foremost.

In 1917 the United States entered the First World War and Russia withdrew from it. At the time the first of these great historical events may have been foreseeable; the second was not. But think now not about the speculation of statesmen but about their entire peoples. In 1917 many Americans, perhaps even the majority of this great people, believed that it was both the power and the destiny of the United States to enter that world war, to decide it, influencing the governing institutions and ideas of much of the world. There were very few Russians, among them even their so-called Bolsheviks and Communists, whose convictions resembled those of Americans. Their Bolshevik or Communist revolution in 1917 was meant for Russia— hopefully for consequences in some of the rest of the world, though that was secondary. In 1917 the United States came into a great European war, deciding it. In 1917 Russia's armies withdrew and Russia retreated from Europe.

Stop for a moment. Please consider if there had been no Communist revolution in Russia in 1917. The American-British-French, etc. armies would have won the war; and their ally Russia, monarchist or not, would have been among the victors. The result may have been some changes in the frontiers of imperial Russia, and Russia would have stayed on as one of the great European powers.

A few years later Woodrow Wilson and Vladimir Lenin died—in 1924, a few days apart. Neither of them was a great statesman. The results were telling. Wilson's idea of "making the world safe for democracy" was already frayed. So was Lenin's international Communism. That failed to establish itself anywhere outside of Russia, whose once-Eastern European empire had been lost.

. . . .

Twenty years after World War I there came a stunning new development. Hitler's Germany went on and on to conquer most, indeed almost all, of Europe. Alone Churchill's England resisted him. Had Hitler conquered England in 1940 would he have proceeded to a war with the United States? I doubt it. Had someone like Hoover been president of the United States in 1940 would he have allied America with Britain in 1940? Roosevelt did. That was his great merit.

On the other side of Eurasia Stalin the Russian leader was willing to go along with Hitler. Still, in June 1941 Hitler invaded Russia. His lust for more conquest and power and his hatred for Communism are still the prevalent explanations for this. Yet those were not his main reasons. Had he conquered at least European Russia (and he came close to it) he thought: what could Churchill (and Roosevelt) do? They would face a Nazi-Eurasian colossus, with a giant German army that would no longer face a war on two fronts—such a world war they could not win. But this did not happen. In December 1941 Hitler did not reach Moscow, and the United States entered the war.

By that time Roosevelt was more and more inclined to believe that his relations with Stalin's Russia were the most important matter. There was a political as well as an ideological element in his thinking. The first was not inconsequential: the United States needed Russia too in the war against Japan. At Yalta Stalin promised him that three months after the end of the war in Europe Russia would go to war against Japan, and so he did. From the Japanese sources we know that in August 1945 the Russian declaration of war against Japan was even more decisive in their surrender than the two American atom bombs dropped on

Japan at the very same time. Roosevelt, like his predecessor Wilson about the League of Nations, put great faith in the United Nations. Even more important was Roosevelt's view of the history of the world. This was that in the development of mankind, now the America of the New Deal represented and marched somewhere in the middle between old England and the rough pioneer Communism of Russia. Many American intellectuals shared this view. But this was not so. Compared to America and England, Russia—its institutions and, even more, the characteristic behavior and the mind-set of its peoples—was not forward and not in the middle but backward indeed.

The "cold war" between America and Russia began soon after 1945. This is not the place to sum up its history during the last 70 years. (One great fortunate condition: all of their geopolitical rivalries notwithstanding, these two superpowers of the world did not go to war with each other.)

. . . .

The main subject of the cold war was Europe, not the rest of the world. Even before the end of the Second World War Roosevelt and the United States allowed Stalin's Russia, at least tacitly, to occupy most of Eastern Europe. (Churchill was most worried about this, though with one possible consolation: the Russians will now swallow almost one half of Europe but they will not be able to digest it—which eventually was to happen.) Immediately after 1945 the United States worried that Communist Russia may now expand into Western Europe, which too would not happen. Even when, in the 1950s and thereafter, popular risings in Poland or Hungary against Communist and Russian rule took place, all of the sympathies of the American people notwithstanding, the United States did not intervene. In a few other places of the world in and after 1945 the triumphant impression of Russian power made local ambitious tyrants proclaim that they were "Communists." A prime example of this has been Cuba, where a loud-mouth such as the anti-American Fidel Castro declared that, given the opposition of Russia to America, he was a "Communist." A prime example of these limitations of the cold war occurred in 1962, when—after considerable hesitation—the Soviet Union was about to ship a few defensive nuclear rockets to Cuba because of the prospect of a potential American invasion of that tropical island. The United States proposed a compromise, and the presence of a few small American naval craft were enough for the Russian ship to turn back.

It was (and perhaps here and there still is) a great and shortsighted mistake to regard the Russian-American cold war as a world contest between Communism and Democracy. (In this respect an ideology of anti-Communism, enthusiastically propagated by many Republicans in the United States after 1947, had produced considerable harm.) In 1951 the aged Churchill told his secretary that by the 1980s there would be no Communism in Eastern Europe. Or indeed perhaps anywhere else. There are hardly any Communists left, not only in Western Europe but throughout much of the world. Communism as a political force, or even as an ideology, is about dead. Instead of being a wave of the future it is no more than a remnant of a short-lived past, its ideology based on a view of human nature that is and always has been wrong (its demise also suggest the eventual ending of the older and still somehow extant categories of "Right" and "Left"—which is another story).

What remains is world-wide nationalism, very much including Russian nationalism, now represented by Putin. His imperial considerations resemble Stalin's; but for Putin, unlike for Stalin in 1945, Communism is a hindrance, not an asset. Russia, especially on its European side, is now bordered by national states fearful of Russian aggressiveness. But in the event of their Russian invasion Putin cannot count on local Communists to support him. Meanwhile his allowance of a primitive Russian capitalism is an element of weakness, not of strength. Still, Russia remains a great power—and Russian-American relations will remain, for a while, as perhaps the principal reality in this world—with which the recent religious-nationalist movements in the Near and Middle East have almost nothing to do.

But meanwhile the very structure of history has changed. Yes, it is still America and Russia and their relations that count foremost. But the relation of states matter less than they have in the past. Popular sentiments matter more than what we used to call "public opinions." The complex and manipulable existence of the former may amount to more even than the existence of unimaginable murderous weapons. But the problems that Putin's Russia may have with its once subjugated European neighbors are entirely different from what the United States has with its neighboring countries and peoples. It is fortunate, too, that—surely on the American side—there is no deep-seated hostility of the American people against Russians. Another providential condition is that America and Russia were and are still distant.

In the Midst of the Crisis of Humanism

One of the problems of "humanism" is the broad (and often also vague) extent of its meaning. "Definitions are tricks for pedants" said Samuel Johnson more than two hundred years ago, and he was one of the wisest humanists in the history of the English-speaking world. One sensible way to define something is to consider its limits, that is: what it is *not*. And even such an evident perspective is imprecise, because of its historical fluctuations: in plain English, the "*when?*" may be as important as the "*what?*"

Looking up the different great historical-etymological dictionaries of most Western European languages one finds that "humanism" and "humanist" are—relatively—late terms, dating back to the fifteenth and early sixteenth centuries. (In England, for instance, one of the meanings of "humanist" as late as 1880 was an ecclesiastic who believed and preached that Jesus Christ was the most important being, but human and not divine. This application is now obsolete.) Still "human" and "humanist" are hardly separable. Yet that, too, is a historically imprecise generalization. There were people in the Middle Ages, who did not call themselves (or perhaps even recognized) that they were "humanists," in the later sense of that term. Moreover, the respect (and often the linguistic and philosophical acceptance) of Greek and Latin writers whose main preoccupation was human thought and human behavior prevailed during the one thousand years after the end of the Roman empire.

But "humanism," as such, was a particularly European phenomenon. Believers in the primacy of human nature and, more important, human thought, existed in many places of the world: but the elevation of a concentrated interest in the unique qualities of human beings rose as a recognizable wave in many fields of life, especially in Western Europe, about six hundred years ago. It was allied with the Renaissance, though not entirely identical with it. The most obvious and visible of the shift of human emphasis was of course in art, especially in its new, emphatic and realistic representations of human bodies and heads.

We may find scattered and remarkable evidences of such in other continents and in other ages: but to speak of the traditions and even of the evidences of, say, Japanese or Visigothic "humanism" makes little or no sense. The "ism," humanism, whether widely used or not, was a "European" phenomenon—arising, perhaps more than coincidentally, at a time when "Europe" and "European" became current terms and adjectives (and when "Europe" also replaced the Mediterranean as the prime theater of history). Gradually thereafter "humanist" and "humanism" became increasingly positive nouns and adjectives, though with limitations. They were infrequent in Eastern Europe, even in countries and nations not under Ottoman rule. Another problem was their frequent absorption by other intellectual features of the so-called Enlightenment. But more significant was the subtle change in the idealization of human qualities that had arisen during the Renaissance. In this sense Machiavelli, with all of his skeptical diagnoses of rulers and ruled, was still a humanist of sorts; but when we read Montaigne or La Rochefoucauld a century after Machiavelli, we find that their fine-penned descriptions of human nature are profoundly realistic: they concentrate on the dualities including the frequent weaknesses of human nature and of the human mind. And within a few decades they are superseded by Pascal, whose burning convictions of human duality flare beyond those of his considerable forerunners. By calling Pascal a humanist we may stretch the term unduly. He may have been less than that—in the then and even now acceptable sense—but he was more than that too. We should regard him as the natural critic of the Enlightenment, rather than of humanism; and one very timely for us, almost four centuries later. "The heart has reasons that reason knows not": the very opposite of Freud and of other twentieth-century thinkers, according to whom the human heart is irrational.

What happened with humanism and humanists is a decline which often becomes the destiny of even the most valuable ideals and illusions. By 1900 the decline of the so-called Modern Age, of the European Age, of the Enlightenment also involved the fading of the humanist ideal. It does not require much historical or literary or artistic knowledge to recognize that a once shining idealism and representation of human nature gave way to a preoccupation with darkness and pessimism and even ugliness representing it. No matter that humanism, even more than the Enlightenment, may have been a principal source of the recognition and legalization and extension of human rights during at least four centuries. "From Dawn to Decadence," so Jacques Barzun wrote before 2000, summed up the intellectual and aesthetic history of the Modern Age—when also the once resplendent words "humanism" and "humanists" had become rare, almost to the extent of their disappearance. Hence this short and incomplete essay: to speak or write of the coming crisis of European humanism is a faint reassertion of something obvious. We are further from the early evidences of its crisis than from its eventual end.

But all is not lost. There are signs that that appearance—and the attraction—of a chastened humanism may be around the corner (or, better put, on the way of its awakening), inseparable as that is from a deeper and sharper recognition of human nature—that is, of human thinking.

During the entire Modern Age there were two achievements of the European spirit that influenced and transformed the thinking of hundreds of millions of people, eventually appearing even in many parts of the world. One was a historical consciousness, a further and deeper spreading of interest in history and its professional as well as amateur study. The other was the ever more universal acceptance and adaptation of the scientific method, again something that was but another phase of the developing ability of mankind to deal with the rest of the world, including even the—so-called—universe. That "science" and its applications transformed much of the world is obvious. What is less obvious is that so much of humaneness, humanity, humanism has been subordinated to it: that, for example, man's history, including that of his psyche, has been made dependent on applications of the "scientific" method, under the—at least partial—illusion of scientific "objectivity." And even less obvious—though beginning to be apparent here and there—are the unavoidable limitations of human objectivity and, consequently, of the desirable limits of a universal "reality." Yet there are signs that shortcomings of the scientific method are, no matter how unceasingly, being felt and sometimes even recognized by all kinds of people. One symptom of this is the slow (though not always conscious) turning inward: a sensing, though perhaps not yet widely recognizing, the inevitable centrality of our minds—and, indeed, of our little globe itself. Already there exist symptoms and evidences of this turning inward. Already more than a century ago the so-called Impressionists started to depict (and even beautify) the world around them with what they saw: that is, not a more and more accurate and perfect representation of "reality" but knowing that this is inseparable from the—often unexpectedly beautiful—conditions of our seeing. And perhaps even more significant: the recognitions of a few thoughtful physicists that, when it comes to subatomic matters, not only our observations but the very presence of these particles being the results of our investigations, of our interference with them.

Thus: a growing, though as yet hardly identified, and not sufficiently conscious, recognition that there is an inevitable connection between the knower and the known. They are not identical; but they are not separable. I am not a prophet but only a man profoundly affected by his sense of the unpredictability of human history. Whether these recognitions will prevail and spread even against or through great coming human and ecological catastrophes is something I cannot even speculate about. But, if not, the chance for a new kind of humanism does exist, and it is likely to emerge in more and more fields of human life—when, among other things, the words "human" and "humane" may acquire a new kind of approbatory meaning of "humanism" and "humanist." If so, with all kinds of consequences, reaching from a slow eventual change in the practices of teaching and learning, or even in an overall recognition of the limitations of the political ideal of popular sovereignty. We (or at least some of us) are already beyond the midst of the crisis of humanism, moving beyond it, but without entirely abandoning our respect for its achievements of the past five or six hundred years, in the

European Age. So the title of his essay, "In the midst of the crisis of humanism" ought really to be "Beyond the crisis of humanism."

I have written about the problematic character of human objectivity and of the scientific method elsewhere and before. But now I insist on one profound matter only. This is that the unavoidable and even inevitable condition of human thinking, of the human mind and spirit, calls us to recognize that the complete and ideally antiseptic separation (the desideratum of "scientific objectivity") of the knower and the known is impossible.

Our knowledge of ourselves and of the world, including of the entire "universe," is personal and participant. A further consequence of this is the—still rare and yet unavoidable—recognition that the universe is "our" universe, which we have invented and keep inventing as much as "discovering"; and therefore, in a very important sense, this earth is the center of our universe, and we the most important, the central living beings in it. The universe "exists" apart from us; but the meaning of its existence does not. There is no arrogance or even shortsightedness in stating this. To the contrary: it is a recognition of humility—a recognition of the inevitable limitations of that otherwise unique instrument, of the human mind. That this recognition, the acceptance of the limitations of the human mind, does not impoverish but enriches us—there are endless examples of this. And such a recognition may open the way to a new, or ever renewed, kind of humanism (one even including a Christian sense of the sinfulness of human nature).

I repeat: an honest and personal and thoughtful recognition of the limits of the human mind does not impoverish but, to the contrary, it may enrich its existential functions:

it enriches the sense of our hearts. But, as almost always, there are contradictory movements in our history. We are still attempting to extend our knowledge (knowledge, rather than understanding) of the universe but, oh, with so much less inspiration and enthusiasm than in the past.

Other divisions will take their places. Democracy will prevail, at least in the sense that it will not be succeeded by a new aristocratic social order. People's ideas move slowly: but there are already some indications that more and more people are concerned with the conditions of our earth, of its possible and potential dangers—including man-made ones. For, whatever "Science" may tell us, our earth is at the center of our universe.

Patriots of earth vs. nationalists of space? Hope not.

INTERNATIONAL? WHAT IT IS NOT

The history of human thought is represented in the history of language. There are words that may be imperfect because human nature is imperfect, though not necessarily misleading. What is essentially misleading and wrong is when a word or phrase is misused: that is—largely—false.

Such is the case of the word, the term, and of the adjective *international*. This is a designation now almost universally employed throughout the globe. It refers, in most cases, to the relations of states, and to those of nations. But during their histories, states and nations have rarely been the same things, and in many ways and places they are not the same things now. It is therefore that their confusion and the misleadingly inflated use of "international" is misleading—and relatively recent.

It was not until the 1860s that this word became current. The Oxford English Dictionary identifies two dates of its first appearance: (a) 1780: "existing relations between different nations" [a rare and infrequent usage then] (b) "the [Marxist] International Working Men's Association, founded in London in 1864." Thereafter: "A person belonging to two different nations, 1870." "Internationalism," 1877: "The principle of community of interests or action between different nations." "Internationalist," 1864. (a) An advocate or believer in internationalism; spec. a member or sympathizer with the International Working Men's Association; (b) One versed in international law; (c) One who takes part in an international contest."

Well we know what happened—more exactly: what did *not* happen—with Marxist internationalism. But something else began to emerge, perhaps as early as the 1860s but certainly not much later. This was the propagation of the idea of national self-determination: among certain English Liberals but soon apparent among Americans, too, regarding it as an inevitable step in the extension of world-wide democracy. Its eventual results were, by and large, disastrous—or so I think and believe. But that does not matter; it is not the subject of this article or essay. What mattered, and what still matters, is the confusion of nations and states, in the minds of people who mattered and who still matter, and with the results of stunning failures. One was Woodrow Wilson's ideal and creation of the League of Nations in 1919. The other, twenty-five years later, was Franklin Roosevelt's idea and creation of the United Nations in 1944–1945. Their failures were already inherent in their misnomers. The League of Nations was an organization not of nations but of states (of governments but not of their governed). So was the United Nations. The first was not set up in the United States but the second indeed in a grand palace in New York. For Franklin Roosevelt this was of principal importance—a crucial step toward the formation of a new peaceful global order. He would give much to the Soviet Union if it would join it. Stalin had no trouble with that: he knew that the United Nations would be a largely powerless forum of the governments of states, not of nations.

The results were obvious. The League of Nations, even without American participation, in the 1930s melted liked a pat of soft margarine in the European frying pan when that was heated up by Hitler. The United Nations, by and large, supported the American decision to condemn North

Korea's attack on South Korea in 1950, but that was about the last and only instance when the UN supported the interests of the free world, including then the United States. Neither the League of Nations nor the United Nations turned out to be anything like real instruments of an "international" order. (Do any of my present readers know the site of the United Nations in New York? I doubt it; nor do I.)

It would be interesting to ascertain when the subject of International Relations first appeared in the curricula of American colleges and universities. It was surely there in the 1920s; here and there probably even earlier. Soon it became an approved undergraduate major; soon thereafter a subject of graduate studies, including the prospect of Masters and then Doctoral degrees in International Relations or International Studies, latest by the 1930s. By the time of the Second World War, and especially after 1945, Institutes of International Studies were founded and established in more and more prestigious American universities, including the renowned Institute of "*Advanced*" International Studies in Johns Hopkins University. There was nothing very wrong with this—except for the persisting confusion of states and nations; whence the persistent inflation of the word "international." The main subjects (and the expertise) of the professors, graduates, and students of International Studies were—and still are—the relations and the characteristics of currently existing states, of their current governments and their sometimes ephemeral politics.

This kind of information about contemporary affairs, without much knowledge of the history (and, consequently, of the tendencies, inclinations, and relations of entire nations) was achievable by almost any serious reader of a reputable newspaper. At best, their professors and students had their genuine interests in history; yet still their main interests and preoccupations concentrated on current events, and on governments rather than on the latter's subjects. In any event, these universities and their International Studies Institutes began to furnish, especially after 1945, experts reputable enough to occupy important positions in the bureaucracy of American foreign relations (such as Rusk or Brzezinski or Kissinger and many others). Generally, their knowledge of world politics was often considerable, while that of their historical conditions of foreign nations were, in most cases, secondary.

This is important: because in an age of advancing democracy of contacts, and connection and reciprocal influences of nations have increased—so that the importances of inter-national (and not only inter-state) relations have been rising, nearly everywhere. But this brings me to an overall essay about the different histories of states and nations: sometimes entangled, or subordinated to each other, but seldom similar let alone identical—which is why the inflated term of International Relations is a misnomer, and will be more and more so.

....

Of course rulers and ruled have always existed. But the idea of a nation appeared only here and there, and not about the same time. Here it is instructive to look at the origin of the word "citizen": a term which is a designation in every state or nation or language throughout the globe. Its origins are, by and large, ancient Greek. "Athenian," "Theban," "Spartan," etc. meant a belonging, with a sense of pride, to a community of a particular and more or less traditional city. (A broader sense of "Greek" also came to exist, but that was more civilizational and cultural and linguistic.) The present usage of "citizen," meaning the legal and acknowledged inhabitant of a state, emerged largely in the eighteenth century, especially in the United States and Great Britain and in France after the latter's 1789 revolution. Thereafter it spread to other nations and languages in Europe and then also elsewhere. But in many languages and nations this connection with "city" does not exist—its equivalent is something like "state-denizen."

Rome was not Greece, and it may be argued that the Roman sense of citizen, "civis romanus sum" (though the "civis" does harken back to "citizen") eventually meant less and less of Roman birth than belonging to an acknowledged (and privileged) member of an empire. This transcended privileges of an urbanity but also of the, as yet, often uncrystallized sense of a nationality or even tribality. Thereafter, at least during the last one thousand years in the history of Europe, states came to exist, more important and sometimes more enduring than their component nationalities. The most evident and graphic example of this were the Germanies, in the middle of the continent. In the Germanies, states—and very different kinds of states—prevailed over their component peoples until, say, 1870. This was, *mutatis mutandis*, so in other parts of Europe too: for example in Hungary, whose founder St. Stephen in 1000 A.D. knew that there were not enough Hungarians to people the otherwise near-perfect geographical shape of his kingdom; he admonished his fellow Hungarians to be hospitable to other nationalities there. In other places and historical situations, too, the sense of belonging to a state at times preceded, at other times was more prevalent than a sense of belonging to a certain nationality. A remarkable example of this has been Switzerland.

In the history of the United States of America in the 1770s the governments of the mother country did not extend some of the self-governing privileges of their citizens to their colonial descendants across the Atlantic. Had the British done so, the declaration of American independence may not have come about then and there; but eventually yes. The conscious existence of nationhood not only preceded but, almost inevitably, led to the establishment of a new and independent state. Not much afterward something similar happened after the defeat and the retreat of the once Spanish and Portuguese colonial empires. A plethora of new states were formed throughout the vastness of Central and South America, from Mexico to Argentina. In retrospect it is amazing how soon, and how relatively easily, the varied inhabitants of, say, Bolivia or Paraguay, not only accepted but recognized and identified themselves as being Bolivians or Paraguayans. It is also amazing how relatively few of these relatively new states fought wars with some of their neighbors along or because of their new and artificial frontiers. (One ludicrous instance was a war between Honduras and El Salvador in 1969, issuing from a riot in a soccer stadium involving their national teams.)

In the twentieth century "national self-determination" had become world-wide. A dozen new national states were formed in Europe (where, however, two of their much-touted examples after World War I, Czechoslovakia and Yugoslavia, broke up into two or more national components twenty-five years later). And at least two dozen new states were formed athwart Africa and also in Asia, most of them after their once colonial rulers had retreated to their mother countries.

Nations were and are more than states. This was the core belief of Adolf Hitler. Early in his life he declared that

he was a nationalist but not a patriot. He had neither loyalty for nor faith in the Austro-Hungarian state where he
was born; his loyalty and faith were attached to Germany,
to Germandom. He dictated that Germany was a people's
commonwealth, a "Volksgemeinschaft," meaning something else and much more than a state. A state was but a
constrained framework, a "Zwangsform." Still he could
not ignore the existence of states beyond his reach (and
even within his empire) during the Second World War. The
nationalist socialism that he proclaimed within and for Germany, he did not impose on all of the states he had conquered in Europe. For Stalin, then, the cause of
international Communism meant little or nothing. What
mattered to him were the dominions and interests of the
Russian state, whether called the Soviet Union or not. Then
in 1989 the Soviet Union broke up because of the sentiments and ambitions of many of their component nationalities.

....

During at least two hundred years nations began to fill up
the frameworks of states. This was, and remains, something
not quite identical with the sentiments of modern nationalism. It was, and still is, a consequence of the rising age of
democracy. But it also meant something else: the rising importance of the relations of nations and not only of states:
of the proper meaning of inter-national relations. The relations of nations, of entire peoples, have been growing. International commerce, international trade have become
near-global. So has international information, including international entertainment. So have a few legal or technical
organizations whose authority may, though only on occasion, supersede the laws and the restrictions of the states.

So has international travelling. But here we may ask: did these developments, and especially the last here mentioned among them, really increase the knowledge that peoples have of each other? Are not their impressions of—and even their interests in—each other hardly more than quick and superficial? History is an inevitable condition (a condition, even more than a circumstance) of the character of every human being, as it is of any nation. Do masses of Americans or Englishmen know, say, France and its peoples better than did some of their ancestors one or two hundred years ago? Perhaps, yes; but also probably, no. Well this may be regrettable: but this too will change. Real inter-national relations will spread and grow. They may not lead to a sense of brotherhood among many people; but of their interdependence, yes. Eventually—and this will not be a rapid development—to the emergence of true inter-national relations. International rules or organizations will not bring about a more-or-less united Europe. That will only come about when many people, including, say, Lithuanians or Portuguese, will both insist and recognize that they are Europeans, that "Europe" may be different from the rest of the globe, that it may even be something like the Switzerland of the globe. That is not at all certain. What is certain is that "international relations" must mean, as they already do, matters both deeper and wider than the relations of the government of states.

THE CRUCIAL SUMMIT:
CHURCHILL AND STALIN

In August 1942 Churchill and Stalin met for the first time. That event was the least discussed and yet perhaps the most important among the many "summits" of the Second World War.

The entire history of World War II proves the then-supreme importance of great national leaders and of their relationships. How contrary this is to the widely accepted and trusted idea: that history and politics and societies are governed by economic and "material factors," that the primary importance of individual persons belongs (if it ever belonged) to earlier ages. The entire history of the Second World War denies this. Its course was set by Adolf Hitler, Benito Mussolini, Winston Churchill, Franklin Roosevelt, and Joseph Stalin. Without Churchill: Hitler may have won. Without Roosevelt: Churchill may not have prevailed. Without Stalin: Churchill and Roosevelt may not have been capable of entirely conquering Hitler.

As the war dragged on, summit followed summit. From 1934 to 1944, Hitler and Mussolini met ten times, but none of their meetings was very consequential—mostly because after 1937 Hitler had the upper hand; Mussolini could not sway him. Churchill and Roosevelt had met first in 1918, neither of them heads of their governments then; Churchill forgot that encounter (this disappointed Roosevelt in 1940). They met twice in 1941, 1942, and 1943, and in 1944 once, almost always in the United States. They sat down together with Stalin two times (the so-called three-man conferences), in 1943 (Tehran) and in 1945 (Yalta). In July 1945, Stalin met Churchill and Roosevelt's successor, Harry Truman, in Potsdam.

The employment of the word "summit" to conferences of heads of state was probably Churchill's choice. It had been, on occasion, applied to a monarch (in 1707 to Queen Anne of Great Britain). There were meetings and conferences of monarchs in the eighteenth, nineteenth, and even the twentieth centuries, and many such gatherings of their chancellors or prime ministers. One of these was the Munich "conference" in September 1938, with Hitler and Mussolini and Neville Chamberlain and Edouard Daladier, the British and French prime ministers, a "summit" where important decisions were made (and of which Churchill disapproved, to say the least). Yet he was often in favor of personal meetings with other heads of government when he thought them useful. (In 1932, Hitler declined a chance to meet Churchill, who a few years later preferred not to meet Hitler.) Unlike in the case of that non-event, Churchill, by and large, trusted his ability to impress other important men. In such instances he was more often right than wrong. That is why his "summits" during the Second World War were important, sometimes dramatic and almost always consequential.

The most important of these meetings between Churchill and Roosevelt were those in June 1942 and January 1943, for these were the meetings that first delayed a second front in Europe (so desperately desired by Stalin), then allowed for an invasion of North Africa that increased the pressure on Germany. In 1942 Churchill was able to convince Roosevelt and the American military leaders that their plan for a landing in western France in November 1942 would be a disaster.

In January 1943, he persuaded them that after eliminating all German (and Italian) forces from North Africa, the Western Allies should go on to invade Sicily and force Italy out of the war. That was the last time the British prime minister had his way with Roosevelt. For many reasons—military, financial, and political—Roosevelt soon was in the position of power. Churchill could no longer convince him and other Americans of his strategic ideas.

Other meetings of heads of state, innumerable nowadays, amounted and amount to largely ceremonial occasions, with agreements and documents signed there but prepared and agreed to before by respective experts. This was not so with Churchill. Of course he brought his experts and advisers with him; but it was he who mattered then and there. This was especially true of his meetings with Stalin. In 1942, as the Russians struggled with very great German forces and the Western Allies suffered stunning defeats, almost everything hinged on Churchill's relations with the then new Russian leader.

Churchill had his faults. But he could think far ahead: he was a visionary about many things. His understanding of history may have been even more profound than his insight into human nature—often the latter was not only inseparable from the former but a result of it. (With Hitler—and Stalin—the opposite seems to have been the case.) But then these two operative nouns may be interchangeable: his *insight* into history and his *understanding* of human nature. Both were high qualities of his mind. And—attempting to assuage Stalin—how he needed them in August 1942!

In December 1941, Churchill's spirits had risen high. The prospect of World War II shone with new colors. The United States had—finally—entered the war, and it was within the same week that a Russian army pushed back a German army a few miles before Moscow. At night on December 7, 1941, Churchill went to bed, relieved. He recalled this within his *Memoirs of the Second World War*. Britain would now live; the Empire would live; Hitler would be conquered; the Japanese would be ground down to dust. Indeed, that was to happen. But it was not to be easy, and the prospects of the war came to seem grim again.

Hitler knew that he could no longer win this world war through the lightning campaigns with which his armies conquered one part of Europe after another—and a large portion of the Russian empire. But he also knew that he was not bound to lose it. Now his entire strategy changed: from short wars to a long war, in which his Germany would prevail, strong enough so that the strange coalition of his enemy powers would sooner or later break apart. Well before that, his armed forces, victoriously, could advance further. And this was so. In the first months of 1942, his submarines disrupted the American and British avenue of the sea across the Atlantic ("A measureless peril," as Churchill once put it). In North Africa, the British suffered defeats, and the German Desert Fox, General Erwin Rommel, came close to Alexandria and Cairo, gates of the entire Near and Middle East. In Russia, Hitler's armies, unlike Napoleon's 103 years before, defied the winter and then resumed their advance, now in the south, reaching the Volga River around Stalingrad and thrusting into the Caucasus.

Defeats, at times amounting to disgrace, clouded the British record and the outlook for the following year. Though America was now side-by-side with Britain, three days after the attack on Pearl Harbor, the two towering British battleships the *Prince of Wales* and the *Repulse* were sunk by Japanese airplanes—all within three hours. Three weeks later, most of Britain's imperial outposts in the Far East were given up. Another six weeks later, Singapore, the crown jewel of the empire in the region, surrendered to a Japanese army half the size of the British and Commonwealth forces there. In June, yet another British attempt to push Rommel back in Libya failed. Worse, immediately thereafter, the considerable military enclave of Tobruk, near Libya's eastern border (and holding almost thirty thousand British and allied troops), caved in, whereafter the German African army entered western Egypt. Even before that, Churchill's wife, Clementine, said to Roosevelt's messenger, Harry Hopkins: "We are indeed walking through the Valley of Humiliation."

Her husband was well aware of that. At the news of Tobruk's surrender he mumbled that defeat was one thing, disgrace another. Earlier that year he mused to confidants: it seemed that British soldiers in World War I were better. The problem was neither equipment nor direction, but morale. That appeared within England and the English as well. The British people's resolute bravery of 1940, when they resisted the German air force's repeated attacks on the United Kingdom, was no longer so dominant: 1942 was not their Finest Hour. What still prevailed was their discipline, rather than their self-confidence. There was no need for public opinion surveys to note this. It involved too the question of Churchill's leadership. Criticisms of that appeared in the newspapers, here and there; and then even in Parliament. A motion for a vote of censure, of no confidence—not in his prime ministership but in his leadership as minister of defense—was brought up on July 1, soon after Churchill's return from Washington. It was defeated by the large margin of 475 to 25.

Above all, acute in Churchill's mind (but also in Roosevelt's) was the question: Would Stalin stay in the war? Would Stalin keep on fighting Hitler, in spite of his dissatisfaction with his Anglo-American allies? Only three years before he had made a deal with Hitler. Would he attempt something like that again? He had some reasons not to trust the Western powers in 1938 when France and Britain signed away part of Czechoslovakia to Germany. He had perhaps even weightier reasons now. More than two hundred Russian divisions were facing and fighting the Germans, while hardly more than six British divisions were skirmishing with a German expeditionary corps in Africa. Stalin also suspected (and rightly so) that yet there would be no second front opening against the Germans in Western Europe, despite his repeated entreaties for British relief. Churchill knew that he had to tell Stalin that the latter would not be coming to ease the pressure by landing in Europe soon. They would focus on the Mediterranean. He had to deliver the news in person. Churchill knew what that was to be like—"carrying a large lump of ice to the North Pole," he said.

And now the days and weeks of July 1942 stretched out, replete with more trouble. A very large convoy of ships sailing for the far north of Russia, packed full of armor and other war material, was disrupted, scattered, and more than two-thirds of its vessels sunk by German attacks issuing from northern Norway. The British now canceled or postponed other convoys bound for Russia for the next few months. This had to be told to Stalin, together with the definite fact that the Allies had decided once and for all that there would be no American-British invasion of western France, no second front, in 1942.

Churchill informed Stalin that he wished to meet him in Russia. He would fly there from Cairo, where he was to confront his concerns about the leadership of the British Eighth Army. He suggested places in the south of Russia, at Stalin's best convenience, to spare both of them a very long journey. Stalin answered that they'd better meet in Moscow. On the last day of that gloom-laden month, Churchill left London. Among other messages, he received a warm bon voyage letter from the king, George VI. "I feel that your visit East will be even more epoch-making than those you have paid to the West." He was right. In his answer Churchill wrote: "in Russia too the materials for a joyous meeting are meagre indeed. Still I may perhaps make the situation less edged." A few days later he dictated a long letter to his wife, from Cairo: "I am not looking forward to this [Russian] part of my mission because I bear so little in my hand, and sympathies so much with those to whom I go." On August 5, he wrote to Roosevelt: "I have a somewhat raw job." He now asked Roosevelt to send Averell Harriman (whom both of them knew well, and who would eventually become American ambassador to Moscow in 1943) with him. Churchill wanted to impress Stalin with the fact that Harriman was coming and would be talking in accord with Roosevelt's wishes.

Years later he recalled: "[During the long flight], I pondered on my mission to this sullen, sinister Bolshevik State I had once tried so hard to strangle at its birth, and which, until Hitler appeared, I had regarded as the mortal foe of civilized freedom. What was it my duty to say to them now?"

That last was a rhetorical question. He knew well what he had to do. Perhaps the questions was not *what* but rather *how*.

"Still . . . it was my duty to tell them the facts personally and have it all out face to face with Stalin rather than trust to telegrams and intermediaries. At least it showed that one cared for their fortunes and understood what their struggle meant to the general war."

His travel was arduous. The aircraft carrying him to Cairo (with a stop at Gibraltar) was uncomfortable, its passengers forced to put on oxygen masks at higher altitudes. His week in Egypt was full of business and action. Rommel was near El Alamein, hardly more than fifty miles from Alexandria, and some of the British Eighth Army was in disarray. Churchill changed its command, summoning Lieutenant-General Bernard Montgomery from England to assume it. He took it upon himself to see as many soldiers as he could. His confidence, in retrospect, amazes: Rommel's army must be destroyed; it will be destroyed. The British forces in Egypt were twice the size of their German enemies, in numbers and in armor, but the Germans, and especially Rommel, had been able to defeat British troops against odds.

Nearing midnight on August 10, Monday, Churchill was flown to Tehran in a now much more comfortable airplane, a new B-24 Liberator commanded by a superb American pilot, William Vanderkloot, fondly remembered many years later. They left Tehran again soon after the following midnight. Churchill brought a less-than-usually-small staff with him. Their plane was delayed and temporarily diverted; Churchill's was not. After a very long flight he landed in Moscow at five in the afternoon on Wednesday, August 12.

He and Harriman descended from the impressive American aircraft. The scene at the Moscow airport was ceremonial (including, among other things, a military parade), more so than for previous Soviet governmental receptions of foreign guests. At least it suggested that, whatever

Stalin's ill humor and frustration with his Western allies, he was impressed with Churchill coming to see him. The prime minister, in turn, was impressed with the lavish details of Russian hospitality, including the comforts of State Villa Number Seven where he and his staff were housed. There were two hitches. He was told to be careful, since in all probability some of the walls harbored secret microphones. (Royal Air Force Air Marshal Arthur Tedder passed a note to the prime minister in French: *"Méfiez-vous"* — be careful.) Churchill then chose to intone a loud tirade denouncing communism and Communists, hoping that the secret-police listeners would get it all. The other matter was less pleasant: as was his custom, he wanted a hot bath after the long weary aerial journey, but instead he was stung by the icy water rushing upon him from a Russian faucet he had thought contained hot. Typical of Churchill, he set his first meeting with Stalin a mere two hours after his arrival in Moscow (and a mere half hour after clambering out of that tub).

Now he was driven to the Kremlin. He and Stalin shook hands. (Others remarked later that, all of his frankness notwithstanding, Stalin rarely looked at the eyes of his conversants.) There ensued their first conference, lasting for more than four hours. It was after midnight that Churchill was brought back to his state villa. He had spoken first, and at considerable length. He began with the nub of the matter. There would be no Anglo-American landing in France in 1942. He reminded Stalin that he had told that to the Russian Foreign Minister Molotov when the latter had been in London two months before. Then he went into considerable detail, explaining why such an invasion would be impossible. There were not enough American troops in Britain yet, not enough armor, not enough shipping and not enough planes for an air umbrella stretching over western France, giving the Allies air superiority. Such details did not impress Stalin. Why, he said, could not the Allies at least land six divisions on the western coast of France? They could, Churchill said, but they could not stay. The Germans would crush them or, at best, force them to leave. And what good would that be? It would surely compromise his and Roosevelt's plans for a serious invasion of Western Europe in 1943. (That there was, as yet, no such definite plan, Churchill did not say.)

Stalin, until then largely silent and glum, now became somewhat impatient, and then sardonic and bitter. Not to risk anything means not to wage war seriously enough. Why were the British afraid of fighting the Germans? He could not demand what his allies would or would not do, "but he was bound to say that he did not agree with my arguments," Churchill wrote. There came a long moment of silence. Then Churchill chose to speak at some length about the now-ever-increasing British and American bombing of Germany. So, Stalin said, there would be no landing in France, and, as Churchill heard the Russian leader, "all we were going to do" was "pay our way by bombing Germany." Churchill—contrary to his temperamental impatience—did not reply. In his hand there was but one good card to play. That was Operation Torch, the planned American-British invasion of French North Africa, less than three months away. Its prospects were infinitely better than an uneasy and temporary descent of a handful of divisions on the western coast of France. Rommel would be struck back at. Franco's Spain would be impressed. Italy would be forced to retreat, eventually from the war itself. The Mediterranean was "the soft underbelly of Europe." (Churchill at one point drew a crocodile, hard and rigid on the top, soft and white on the bottom.) All of this would benefit not only the prospects for the war but Russia itself. With the Mediterranean eventually taken from the Axis, the Allies would have easy access to the Black Sea to help the Russians (when and if needed), and for the Russians, a better route to open waters.

Stalin seemed to pick up on the prospects of Operation Torch remarkably quickly and well. Whether he had known something of these Allied plans (mostly through American contacts and intelligence agents), we cannot be sure. He may have had an inkling but nothing that was certain. Churchill, who was impressed with the directness of Stalin's reaction, was gratified and relieved. He now asked Stalin to meet with him again: Stalin said of course. So that momentous day— and some of its succeeding night—had passed.

Churchill did not retire to bed before two in the morning. He had now been up for something like twenty-six hours, interrupted by snatches of sleep during the long flight from Tehran to Moscow. Understandably he rose the next morning—Thursday, August 13—somewhat later than was his usual wont. Before him lay the prospect of a (relatively) restful day. He proposed to call on Stalin at ten that night. Stalin suggested eleven. The working day began with Churchill meeting Molotov, the latter expressionless and slab-faced as usual. Then Churchill lunched and rested in his state villa during an undemanding afternoon, inspecting the garden with its fountains and goldfish which inspired his interest, and then a sumptuous air-raid shelter which did not. Late at night he was driven to the Kremlin, where only Stalin and Molotov were waiting for him, with an interpreter.

There began a very disagreeable discussion, indeed the low point of Churchill's four days in Moscow. Churchill suspected that his and Stalin's amiable good-bye the previous evening was not the end of the matter. He warned Molotov at noon: "Stalin will make a great mistake to treat us roughly when we have come so far." (Molotov: "I will tell him what you say.") Did Churchill know the old Russian practice of bargaining, which was to impress or stun or shock the opposite side by stating the maximum Russian aims or demands soon after the beginning of the conference and then negotiate some kind of compromise later? Certainly this session began with Stalin: harsh and demanding. He produced a document, handing it to Churchill, who said that he would read, study, and respond to it later. The memorandum employed a standard Soviet habit of stating things that were at least arguable with their habitual phrases of certainty: "As is well known . . ." "It is known . . ." And so on. "It is easy to grasp that the refusal of the Government of Great Britain to create a Second Front in 1942 in Europe inflicts a mortal blow to the whole of Soviet public opinion." The gist: "We are of the opinion therefore that it is particularly in 1942 that the creation of a Second Front in Europe is possible and should be effective. I was however unfortunately unsuccessful in convincing Mr. Prime Minister of Great Britain thereof." Even while this was being translated, Churchill broke in to say that he would answer it in writing, but also that the British (and American) decisions had been taken, and that there was no use arguing about them now. This did not deter Stalin, who continued to insist that the making of a second front in western France was possible; that the British were going back on their earlier promises; and that they, unlike the Russians, were afraid to fight the Germans.

During these two hours Churchill gave as good as he got, which impressed the Russian, who proposed a dinner meeting the next night, August 14. Churchill said that he had planned to leave early in the morning on the fifteenth. Stalin was stunned by this, and proposed that the prime minister remain a day longer. Churchill said yes, but that there ought to be a real spirit of reciprocal understanding and of the appreciation of their mutual alliance by the Russians. Stalin now retreated somewhat, and they went on talking about some military details. Then Churchill began asking Stalin about the Caucasus, where loomed the prospect of a German descent into the Near and Middle East, with the direst of consequences. Stalin said that the Caucasus would be held. Thereupon, Churchill, assisted by Harriman, suggested that British and American air units could be ferried to Russia if needed, to south of the Caucasus but also to Siberia over Vladivostok. Then Stalin made one last sardonic remark: "Wars are not won with plans." But when they left, Stalin put his arm out and gave Churchill a warm handshake.

That was another long day for the latter. Churchill had a long and good night's sleep. Then he composed his answer to Stalin's memorandum, and drafted other telegrams and letters to the War Cabinet and to Roosevelt. He lunched with General Alan Brooke, the chief of the Imperial General Staff. He had a headache, but then he rested. He was taken to the official dinner in the Kremlin, at eight in the evening.

This was a ceremonious event including almost one hundred people. Churchill was seated at Stalin's right, Harriman at Stalin's left. Stalin was amiable, at times jovial. Their conversations were somewhat hampered; the English of Stalin's interpreter, Vladimir Pavlov, was far from perfect (which was not so with the Russian of Churchill's now summoned translator, Major Arthur Birse). Among other matters, Stalin attempted to please Churchill with a recollection, true or not. He said that when many years before the playwright George Bernard Shaw and MP Lady Nancy Astor visited Moscow, the name of Churchill came up. The latter had said that Churchill was "finished," whereupon Stalin had said (or said that he had said) that in a great crisis, the English people might turn to Churchill again. Lady Astor also said that the prime promoter of Allied intervention against the Bolsheviks in the Russian civil war of 1918–21 was not then Prime Minister David Lloyd George but Churchill. That was so, Churchill agreed. As the prime minister recounted in his memoirs, Stalin "smiled amicably, so I said, 'Have you forgiven me?' 'Premier Stalin, he say,' said Interpreter Pavlov, 'all that is in the past, and the past belongs to God.'" That was one of many occasions when Stalin (the once Orthodox seminarian and now supposedly atheist Communist) invoked God during the four days of talking with Churchill, who thought that remarkable.

That dinner stretched on and on. Churchill was tired but also angry. He was impatient with the endless sequence of toasts and what he thought was Stalin's insufficient attention to him. It was now beyond one in the morning, with a film about to be shown. Abruptly Churchill got up and then marched down a long corridor with his staff toward the exit. Stalin unexpectedly jumped up and trotted after them. Catching up with Churchill finally, they shook hands. (He also said their differences were but disagreements about methods.)

The next day, Saturday the fifteenth, was taken up by long conferences between the Soviet and British military authorities. They were largely inconsequential, the Russians again complaining about an as-yet-nonexistent second front. The British were still worried about whether the Russians could hold on to the Caucasus. Churchill was a tad more confident about this than was Brooke. At seven he was driven to the Kremlin to say good-bye to Stalin. He brought up the Caucasus again. Stalin was reassuring. They talked for an hour or less. Churchill was about to leave; he would fly off very early the next morning. Suddenly Stalin moved close to Churchill and proposed that they repair to his home for drinks. Churchill assented. They went across the empty Kremlin courtyards to Stalin's apartment. There appeared Stalin's daughter, briefly; and then a substantial dinner, including a roasted suckling pig which, not at all briefly, Stalin enjoyed. They now stayed together for another six hours, talking about a great variety of things. These included Stalin's account of how difficult (and inevitable) was the forcing of the Russian peasantry into collective farms. Churchill brought up Operation Jupiter, another second-front plan involving a British invasion, perhaps with Soviet support, of northernmost Norway (this was a pet plan of his for at least another two years, scotched by the chief of the Imperial General Staff and other military and naval authorities). Before Churchill departed they read and approved a joint communiqué. Churchill came back to his residence after three in the morning on Sunday, August 16. He was to be driven to the airport less than two hours later.

Then he slept in the plane for long hours, flying over the vast gray and green tracts and fields and steppes of central and eastern Russia. When he woke up they were almost beyond the Caspian Sea, soon beginning their descent to Tehran. There, refreshed in the cool residence of the British Legation, he wrote a cordial thank-you message to Stalin, and two long accounts to Roosevelt and the War Cabinet.

His four days and nights in Moscow now slipped into the past, which, at least in the words of Stalin, "belonged to God."

Eight years after his journey to Moscow, Churchill wrote its lengthy and fairly detailed story in his history of World War II. In his account of August 1942 he suggested that this was the crucial "summit," perhaps the most important of his five encounters with Stalin during the war. We may gather this from the very extent he chose to devote to it: almost thirty pages and two chapters, longer than any other description of his wartime meetings, except for the three-man conferences in Tehran and Yalta. They are not contradicted by other sources, including the records of those who came with him to Moscow, by his leading wartime adviser Alexander Cadogan, General Brooke, Major Birse and Churchill's personal physician, Lord Moran. It is telling to compare the two entries in the diaries of the reserved and skeptical Brooke, who on August 13, 1942, wrote that Churchill and Stalin "are poles apart as human beings . . . [Churchill] appealed to sentiments in Stalin which I do not think exist there." Yet he later wrote about the visit: "looking back on it I feel that it fulfilled a very useful purpose, that of creating the beginnings of a strange understanding between Winston and Stalin."

The most important, nay, the decisive matter we can state about Churchill in Moscow: it is impossible to imagine that any other British public figure could have appeared in Moscow and impressed Stalin as he did. For Stalin the dominant matter was Churchill's character. So much so that it may be easier to grasp what the secretive and crafty Stalin thought of Churchill than what the voluble and sometimes loquacious Churchill thought of the Russian. The latter had some respect for Churchill before their first person-to-person meeting, but then this congealed rapidly during and after that initial encounter. Whatever the British armed forces did or did not do, Churchill was a fighter. And beneath all the ceremony of those days and nights ran the slow, steady, crude flow of Stalin's temperament: his despising weakness of any kind, and his respecting (and even admiring) strength—including in those who stood up to him.

Churchill's assessment of Stalin, on the other hand, was primarily historical. Hence his seemingly (but, at least to me, only seemingly) contradictory views, perceptions and opinions of the man. They were sentimental, too; but also profound. Churchill is often accused of a certain hypocrisy: he needed Stalin in the war, so he was more than willing to overlook the brutalities of this Communist dictator. In other words, the Churchill who argued and fought so bitterly against appeasing Hitler was ready to go long miles in the company of Stalin. But Hitler—for Britain and Churchill (indeed, for Western civilization)—was more of a threat than Stalin. Hitler was ready to rule all of Europe; Stalin, the eastern half—and half of Europe was better than none. That was the essence, and the consistency, of Churchill's historic and strategic vision.

Russia had been Russia, Russia was Russia, Russia will be Russia and that was that. Here was the difference between Churchill's and Roosevelt's views about Stalin's Soviet Union. The American president saw it as a rough, pioneer empire, thinking that eventually Russian and American democracy might be reconcilable. Churchill saw and thought otherwise. What the Soviet Union represented was not a rough, pioneer experiment toward the mass democratization of the world. Russia was not forward, it was backward. Its strength welled up not from its vision of the future but from the atavistic tradition of its people. Churchill was convinced of that, sometimes perhaps even too much so. Hence some of his speaking and writing about Stalin in flowery phrases on occasion, attributing to Stalin a wisdom that Churchill hoped would govern his war leadership and their alliance. That wisdom was there. But it was not enough to govern Stalin's motives—and fears.

RE-READING "MADAME BOVARY"

My paperback copy of *Madame Bovary* (I have another copy too) has four appendages, none of them necessary. At its end there is the full record of Flaubert's trial: the prosecution in 1857 condemning the novel for its immorality; there is the excellent speech for the defense by Flaubert's attorney; there is the short verdict acquitting Flaubert and his publishers. At the beginning of my book, there is an introduction by Mary McCarthy. She despises and dismisses the woman Emma Bovary. Reading this was one of the impulses that made me write this.

I do not despise Emma Bovary. I also have a great respect for Gustave Flaubert. I think moreover that now, almost 200 years after its publication, Emma's story and Flaubert's description of it are not as simple as they have been read and seen during much of those two centuries. Let me begin with Flaubert before I turn to Emma. The standard epithet has been that "Madame Bovary" was the first "realist novel," and in this respect a new kind of masterpiece. There is some truth in this. Flaubert thoroughly understood that the novel is not a prose form of epic literature but something quite different. Unlike an epic, it must deal with "real" people. In "Madame Bovary" its subjects are, almost entirely, people of the French provincial middle class: bourgeois, but of an order somewhat lower than some of the large inchoate group ruling much of France in the nineteenth century. This was something—not entirely, but almost—new at the time.

But more important is Flaubert's preoccupation with, or call it his concentration on, what is "juste," of which its English translation ("just," "exact") is not really a perfect equivalent; the closest and best English adjective would be: "right." It is the essence of Flaubert's "realism." We know that fairly early in his life he convinced himself that one must write about something that one knows a great deal about. (After "Madame Bovary" he departed from this at least twice during his authorship, in "Salammbô" and in his "Temptations if Saint Anthony"; but then he returned to it twenty years after "Madame Bovary" in "Sentimental Education" which, together with "Bovary," is one of his two enduring masterpieces.) Whence, too, his meticulous, and nearly obsessive concern with the "right" word, the *mot juste* that, beyond accuracy, even forms his style from page to page in a somber major key. We also know that he was much interested to know more and more about an adulterous provincial woman, also in Normandy. He was critical (not the "mot juste") of her as he was to be of Emma Bovary.

But did he despise Emma? Yes: and no. This is what interests me now, two hundred years later. Also: Emma Bovary was a period piece. But so were the characters of Cervantes— or Swift—or Jane Austen. That does not matter. What matters is "Madame Bovary"'s durable relevance for its readers today, perhaps especially for women readers. I see a duality here. On the one hand Emma Bovary was the captive of a world that no longer prevails. Her marriage was dismal, and she was devoid of freedom. Divorce was not absolutely impossible then, but seemingly impossible for her. Now things—circumstances, conditions, choices, and even desires and sentiments—have changed. Women have freedom, they fashion and try to govern their lives accordingly, even when they may or may not recognize that freedom is a task, that in so many ways it is more difficult to be free than not to be free. Their lives, their problems, even their loves are not like Emma's. But on another, perhaps deeper womanly level, they will understand Emma; and I doubt that, unlike Mary McCarthy, they despise her. They may understand her passions—rare though such passions might be in the world now. "Rare" meaning "cool"; or "rare" meaning "infrequent"? Depends on the reader. I cannot tell.

But whether now, or two hundred years ago: Emma's life is an almost sordid tragedy; and any attentive reader will sense early what may happen to her in the end. She is a romantic. Her undoing is the result, unbounded, of the first. Here is one evidence why Flaubert's condemnation for having written an immoral book is absolute nonsense—and not only because of our viewpoints or standards two hundred years later. Flaubert about the dénouement, her life's peak, Rodolphe making love to her for the first time. ". . . with a long shudder and her face hidden, she gave herself to him." *She gave herself to him*: five words, nothing more. Surely it is not like "Lady Chatterley's Lover."

Of course Rodolphe deserts her. She is ready to flee with him, anywhere. He is not. After that, there may be one weakness in Flaubert's creation: Emma's depression, her "brain fever" lasting forty-three days. More important is her second tragedy. Her second love affair, with Léon, is a desperate and weaker repetition of the first. He too deserts her but there is less of a drama than there was with Rodolphe, who was a cad, while Léon is something of a weakling. But now it is less her longing than her existential situation that seems hopeless. She has amassed debts she cannot pay. Flaubert's account of her descent into hell is now stark and rapid. She poisons herself with arsenic. Her suffering and death are awful.

But that is not the end of the book. There follows a ten or twelve-pages-long last chapter, whose main figures are her badly stricken husband Charles and the pharmacist Homais to whom I must yet return. Then rings the death-knell on the very short last page. Charles falls to the ground, dead. His small daughter, Berthe, has no one left to sustain her. She disappears into poverty. The last walks together with her father are among the saddest things I have ever read.

Throughout "Madame Bovary" Flaubert disparages Emma, and not only because of her extreme passions: among other things, because of her occasional neglect of her daughter. But did he despise her? Was she not a human being—and significant enough to be the subject of an entire (and new) novel? There is a passage near the story's end. Her dolt of a husband, whom she disliked and sometimes even despised, is sobbing with his head against the edge of her sickbed:

"Don't cry" she said. "Soon I won't torment you any longer."

"Why? What made you do it?"

"I had to, my dear," she answered.

"Weren't you happy? Is it my fault? But I did everything I could."

"Yes, I know. You are very good." And she caressed his hair slowly.

Gustave Flaubert did not despise Emma Bovary. The main object of his contempt was someone else, to whom I must now turn, in order to say something about Flaubert's view of the world at that time.

....

The main object of his anger is not Emma but the pharmacist Homais—a principal figure in "Madame Bovary." His book ends not with Emma's death, but with a long chapter, mostly about Homais, and it ends with Homais. "He has an enormous clientele. The authorities cultivate him and public opinion protects him."

"He has just received the Legion of Honor."

This is the last sentence of "Madame Bovary."

Flaubert despises Homais, but not only because this pharmacist is a prototypical bourgeois. Homais is even more pompous than stupid. Unlike most of the provincial bourgeois in the 1850s, Homais keeps spouting the clichés and the ideology of the once Enlightenment: Voltaire and his kind, larded with Latin medical phrases as often as he can. He is an endless bore throughout: an Atheist and a Scientist and a Progressive. The Abbé Bournisien, the priest who attends Emma at the end, is Homais's opposite. He is gentle, kind, and ineffectual. Flaubert knows his intellectual weaknesses, yet Abbé Bournisien is humane, understanding, forgiving. But he is a minor figure in this book where, as we have seen, Homais is not.

This brings me to a consideration of Gustave Flaubert's place not only in the history of literature but perhaps in the history of thought. Yes: he disliked the bourgeois; he was a committed "realist." Other romantic and then "realist" writers throughout the nineteenth century despised (or at least claimed to despise) the bourgeois, even when the latter had made it possible for them to profit from literature. But Flaubert was not quite like them. For one thing: his concerns were often deeper than were theirs. He was obsessed with language, with the words and phrases not always used and spread by the bourgeois, but also by some intellectuals and artists. He was incensed by the acceptances and repetitions of clichés, appearing within the otherwise subtle and refined French language. Near the end of life he wrote the (unfinished) extraordinary "Bouvard and Pécuchet," a compendium of irony, consisting of the conversation of two bourgeois readers and "thinkers," not only larded with but chock-full of clichés, ergo: a mosaic or even a panorama of near nonsense. Flaubert died before finishing "Bouvard and Pécuchet" and also while he was still limning his "Dictionary of Accepted Ideas" that he thought might accompany B. and P. Perhaps fortunately it did not: for his "dictionary" (which he had begun to compile many years earlier) lists clichés that are cryptic rather than telling. And he knew very well that the rightness of a word is not determined by a dictionary. It issues from within a speaker or a writer. Flaubert's obsession was to detect and demonstrate false words and phrases—whence his compulsion for finding his "mots justes."

But allow me to argue something more important. Perhaps Flaubert ought not to be ranked as one of the principal writers of "realist" literature. How different he was from, say, Zola: that much we may see and know. But he was also different from Stendhal or Maupassant, because in an important sense he was ahead of them. He, too, despised the ideology of the bourgeois: but he was also skeptical of the ideas of Progress, and of the Enlightenment. We can read this throughout "Madame Bovary." He saw the shallowness of Voltaire or Diderot and of their kind. So he was a forerunner of a slow movement, beginning in the second half of the nineteenth century, and emerging only here and there, slowly, through the twentieth (that slowness of ideas in the age of democracy that Tocqueville foresaw)—the inevitable limitations of a scientism anchored in the belief that human knowledge is unlimited and boundless—and that this must be now deepened not by an increasingly blind belief in its boundlessness, but enriched by a recognition of its limits.

Thirty years after Flaubert finished "Madame Bovary" Tolstoy published "Anna Karenina." It has been often classified as *the* other great nineteenth-century novel, dealing with passion, adultery, and death. Both Emma and Anna killed themselves. It has also been said (by Mary McCarthy too) that Anna was a tragic heroine, but Emma was of course not. This is not true—or, rather: not true enough. Both seem to have deserved their respective fates; but Tolstoy had no pity for Anna. Instead of Homais, Tolstoy invented Levin, a likeable and eventually happy, married man. "Happy families are all alike; every unhappy family is unhappy in its own way." This is the very last sentence of "Anna Karenina"; and it may be argued that an opposite generalization may be true.

"Madame Bovary" (also two-thirds shorter than "Anna Karenina") is the greater book: perhaps because the anti-bourgeois and anti-romantic Flaubert was also a profound romantic, despite himself, after all.

THE CULT OF ONESELF

Benjamin Constant's *Adolphe* was published in 1814, more than two hundred years ago. He had written most of it seven years earlier. It was often reprinted later. Many people liked it and respected it ever since. In 1814 and 1815 some of them saw it as a minor sensation, something quite new. Often it had been categorized as an autobiographical novel, or as an early exemplar of French romanticism, or of a new genre, a "culte de moi," a cult of oneself. The purpose of this essay is to argue that none of these labels is quite right.

Benjamin Constant was a very good writer, while not a very admirable or even attractive man. Sometimes this is so with significant artists. But when it comes to a writer, and especially to a writer of remarkable prose, to separate the qualities of his writing from those of his character is never entirely possible. In Constant's case, and especially in "Adolphe," this may be impossible. An autobiography calls for a writer's interest in himself (rather than in his times: recently the subject of so many autobiographies is history—the writer's life and then his thoughts about his times). But two and more centuries ago one of the consequences of the Enlightenment was a deepening of interest in human knowledge—so that a man or a woman should now attempt to come closer and closer to describe reasonably what was happening in their hearts. Well—some autobiographers tried that in one way or another but, by and large, they listened not to Pascal's great seventeenth-century dictum, that the heart has reasons that reason knows not. A proof of this is "Adolphe" that Constant wrote when he was forty years old. Well after its publication some saw in it a new kind of writing: The Cult of Oneself. Well—with all of his individualism this did not really happen. Benjamin's contemporary Edmund Burke once wrote that law sharpens the mind only to narrow it—and much of this can be true pf autobiographies too. In "Adolphe" Constant was narrow, and sharp, because of his utter self-regard. He thought, or at least pretended to have gone deep; but two centuries later this depth does not impress us, because whatever he illuminated was his "persona"—which is not quite the same as "himself."

He pretended that his heart had dominated his reason till the end. That was not so. "Adolphe" consists of two human beings, Adolphe and Ellénore. Throughout his life Constant (his real name, not an omen) was inconstant, in love with love, just about always with women who were older than he. Nothing wrong with this. Ellénore in "Adolphe" is the third or fourth of his great love affairs. But in real life she did not exist. In some ways (but only in some ways) the Benjamin-Ellénore tragic love affair has much to do with Benjamin's contemporaneous parallel relationship with Madame de Stael (also older than Benjamin, though not by much), who had a good estimation of Benjamin's mind but really did not care very much for him. "Adolphe" then is an account of what a not very attractive man (among other things, Benjamin was physically unattractive too) may be capable. Ellénore is a very, very beautiful woman, a once refugee from Poland, protégée and mistress of the Count de P., an honorable man with whom she had two children. She is also good-natured and honest, though not particularly intellectual. Benjamin decides to fall in love with her; she, decently and honestly, shies clear of his advances—until his repeated presence and frequent declarations created a response in her heart.

"At last she gave herself to me without reservation." To this sentence I shall return. Most of the rest—which is the major part of the book—describes "Adolphe" or, rather, the turbulence of his mind, since he will no longer be in love with Ellénore. (What "in love" means is described in long paragraphs interspersed throughout this book, here and there worth reading.) But he is not capable of breaking with Ellénore, who knows this well, and suffers dreadfully because of it. His passion for her declines, or is transmuted into pity. They wander across Europe; at the end she dies, wasted and ill, in misery. Many women (and also some men) who read "Adolphe" two centuries ago had tears in their eyes.

"At last she gave herself to me." Let us pay some respect to the nineteenth-century literary manners. Descriptions of physical love-making were rare then, even for a writer such as Maupassant, who wrote many hundreds of short stories dealing with sensuous and ambitious women, but never a word about the climax of each story, about the moments of seduction (in Maupassant's stories often it is not a man but a sinuous crafty woman who is the architect of the seduction). Yet in "Adolphe" this short sentence of seven words will not do. After more than forty pages of his wooing Ellénore, interspersed with various passionate summaries of what love consists of, there is no word about a first kiss or caress or her gradual acceptance of his ardor. But in another paragraph after Ellénore's surrender to him: "I walked with pride against men . . . The very air I breathed gave me pleasure, to thank her for the unhoped for, the immense blessing she had deigned to grant me." This recitation of a temporary bliss is not very convincing.

Very soon there are signs that Adolphe was getting tired of Ellénore. Or, more precisely: their relationship has become reversed. At first Adolphe was passionate and Ellénore reluctant; now the opposite was happening. When it comes to a love affair, that is not unusual, alas: but Constant spends more than half of his book on Ellénore's decline and fall. She abandons her protector, even her children. Adolphe does not abandon her, but he no longer loves her—worse: he even lets her know that. His passion for her has passed down to pity, all of which he describes in probably unnecessary detail. In this case pity may be an alternative, a substitution for true love: Constant describes himself as generous, but he was only weak.

In "Adolphe" he eschewed descriptions of physical love. His readers understood this at the time, and we may respect them for that. What is strange for us to understand is how (both before and after its publication) he read "Adolphe" aloud for many hours to sophisticated listeners whose attentions were eager and rapt. Two hundred years later our span of attention has become extremely limited; but, except for the cheapest kinds of writings, this is not why, unlike in the past, some people will read detailed descriptions of lovemakings. Passion was a fine thing two hundred years ago and it still is: but its minute analysis is not. But this short essay is not a historian's attempt to reconstruct sensitivities and even manners. What strikes me in "Adolphe" is not that Constant was untrue to himself—but that, in more than one sense he reversed Pascal's immortal phrase: his mind had reasons that his heart did not admit.

In this respect he was much less a romantic than a late product of the Enlightenment. His virtue was, and remains, that his writings are extraordinarily clear, including his political ones. But while he did not intend to deceive his audiences and readers he often deceived himself. One example: in his lucid "Red Notebook," a shining account of the first twenty-two years of his life, he wrote that in London many people took him for a Scot, "since I had retained the Scotch accent from my earlier education in Scotland"—nonsense, since he had spent hardly more than a year in Edinburgh. He *was* a classic liberal, admired by many people for his political writings, even though he had changed camps more than once. He wrote that Napoleon, returning from Waterloo, spent three hours talking with him in the gardens of Elysée. I find this unlikely, but I do not know. He may have deceived some people; he also deceived himself. And yet he was authentic. Such are the complexities of human nature.

A Faint Reason for Optimism

When I was very young in Hungary, I read Oswald Spengler's "Decline of the West," of which the original title in German is even bleaker: "Untergang des Abendlandes": the Sinking, or the Foundering of the West. Many people, at least in Central Europe, had read that ponderous (in more than one sense) book by then. I was depressed as well as impressed by it, or so I remember, though it does not seem to have left a definite memory in my mind. I had read many books and articles written about the inevitable sinking of the West, even in generally optimistic America, where Spengler, because of his heavy German rigidity, had had not much influence. I am not preoccupied by Spengler but about the state and the future of my greater homeland, the "West." (I am, and was, and will never be a prophet; I am a historian, at best (and a somewhat shaky "best") a very occasional prophet of the past.) And I am writing this short essay not about the future but about some matters in the still present and more and more recent past.

Yes, we are at the end of an age: and our knowledge and sense of this is—relatively—new. At the end of the Middle Ages many people sensed the looming presence of great changes, but no one called that "The Passing of the Middle Ages" (a great book with that title and theme was written by the Dutch historian Johan Huizinga, published in 1920). Our recognition of the ending of an age has been the result of the growth and of the presence of our historical consciousness, one of the great achievements of the European Age, together with the, at times hardly imaginable, development of "science," that is, the science of nature—which, however, is part of history and not the other way around.

That now past and passing age may be called the European Age: better and more precise than the indefinite and still current "Modern Age." More than six centuries ago some Europeans started to travel around the globe and then their descendants began to settle parts of it. About five or six centuries ago the most powerful kingdoms and states of Europe were the Spanish and the French and the English, establishing their colonies around the world. After the Second World War they no longer were. They abandoned their colonies. The domination of white people and of "Europe" came to an end. White people within Europe too now had smaller families and fewer children than before. In the United States too the numbers and the presence of Central and South American people were growing. So did, by and large, the life expectancy and the living standards of almost all people of the globe. To sum up, or merely to list the retreat and the diminution of the white race would fill the contents of a small book. So would a book about the degeneration of civilization and of much of the culture within Europe and America, even without arguing or explaining the reasons for these melancholy developments.

But: are they inseparable, indeed identical, with the fatal decline of the "West"? Yes: and no. Or rather: a large yes, and a smaller no. The "West": is it declining and diminishing? Yes, but also no. One symptom of the "no" is the present ubiquity of the English (or Anglo-American) language throughout more and more of the entire world.

And an even larger development has been the growth of "democracy"—a word that had dubious and skeptical meaning even less than two hundred years ago when a solitary French nobleman, Tocqueville, saw something much larger and deeper than the great historic periods of Europe, indeed of the white race, from Greece and Rome and the Middle Ages to the "Modern" or European Age. There was an "aristocratic" age, its regimes differing here and there from time to time, but essentially ruled by minorities, of small portions of entire peoples. Now something else was happening and developing: rule by majorities—or at least in the name and with the consent, of majorities. Tocqueville saw this in the United States, a great state and nation and people then almost new. The title of his great and still in many ways relevant book of two volumes, "Democracy in America," is telling. Contrary to some accepted beliefs, it is a book (and especially its second volume) about democracy even more than about the United States. What was established and what was happening in the United States in the eighteen-thirties would, he thought, come about, in various ways, elsewhere in the world too, and perhaps, especially in Western Europe. He did not speculate or write much about the latter. His preoccupation was the—to him inevitable—rise of Democracy. And now let me ask: the near-universal rise of democracy and the near-universal decline of "the West": are they inseparable? Or synonymous? This is the theme of this brief essay, written by an old man, almost two hundred years after Tocqueville, whom he has read and read and admired through a lifetime.

I think that a universal or general history of "Democracy" has not yet been written. Of course there were episodes of democratic rule in Greece and Rome; here and there toward the end of the Middle Ages; here and there among primitive peoples. Of course democracy and populism and equality and the sovereignty of the people were, and are not the same things. But the rule of democracy—as we know it, and want to protect and preserve it—was the creation of the West. There were old democratic elements within the more or less Scandinavian kingdoms. Their seeds were there in some of the various Western European and English parliaments in the Middle Ages, consultative assemblies under the rule of kings, as yet devoid of equality or people's sovereignty. There were houses of lords, of one kind or another. But, especially in England, the respect and the role of the parliament, especially that of the House of Commons, widened; and, with that, freedom broadened slowly down. So, in any history of democracy, what took place in England in the seventeenth century may be as telling as what took place in France one hundred and fifty years later. A King had been condemned to death by a (truncated) Parliament. A large crowd of Londoners groaned in 1649 when they saw the head of their King Charles axed off. For the next eleven years England became a republic of sorts, called a Commonwealth—an English word for a *res publica*.

The word "republic" was not popular among Englishmen (as it would be in France and elsewhere 150 years later). Nor was the term "democracy" current, except here and there. It was not derogatory, but not yet frequent. (One example: after most of the Presbyterian party was expelled from Parliament, they called the commonwealth "a heretical democracy.") The history of words does not only reflect: it signifies history, the development of thought. What is so telling about England almost four hundred years ago, including their Civil War, is the public language then already current. "Democracy" only here and there; but "the sovereignty of the people" and "the freedom of the people" almost everywhere (except of course by the Royalists).[1] The great poet John Milton, who despised the King, kept recalling "old English fortitude and love of freedom." Universal suffrage not yet: but it was edging closer. It had its vocal and often popular advocates (such as the Levellers, etc., forerunners of social democracy, perhaps even of some kind of communism: "Unless we that are poor have some of the land to live upon as freely as the gentry it cannot be a free commonwealth."). That kind of popular sovereignty was not espoused by Cromwell. But by some of the Parliaments: yes, for nearly twenty years they depended on the army; but also on the respect with which so many Englishmen still treated Parliament. Throughout his career Cromwell wanted to keep it. For five years he was something like a near-constitutional monarch: the Lord Protector of England. In 1658 he died. The weak son of the beheaded king came back to England, supported then by the army, and a very large majority of the people—without more bloodshed, without a renewed civil war. A restored but no longer absolute monarchy, a constitutional one, King and Parliament—the rule in England ever since.

Cromwell had no followers in Europe. (Nearly four hundred years later I respect him for many things: but I would have been—and perhaps still am—a royalist, admiring Charles's bravery during his trial and his tragedy.) So: whence my emphasis on England, perhaps a paradigm of a coming democracy, yet uninspiring and unrespected elsewhere in Europe then, unlike the French Revolution a

---

1 The history of words: enemies of Cromwell in a pamphlet called him "a certain mechanic fellow." For about two centuries thereafter "mechanic" was often a derogatory adjective.

century and a half later—a brief episode in England's long history, an interruption of its monarchy for a few years near four hundred years ago? Because of America. Because of the United States. Primarily so. Most of the people who came to North America during the seventeenth and eighteenth centuries were Englishmen, Scotsmen, Irishmen. They knew little about Cromwell. But most of them knew English popular customs, English popular liberties, English speeches, Protestant and even independent religions. (One exception to the latter were many Irish immigrants, and the founders of the colony of Maryland.) A century later the North American declaration and struggle for independence, for the sovereignty of the people, and the very elements and structure of the constitution of the United States were (and are) unimaginable without recognizing the American political and social and religious customs and institutions and language that had begun to spring forth in seventeenth-century England.

In 1789 most American public speakers thought and spoke of the United States as a republic, rather than that of a democracy. At that time, and for more than a century thereafter, the adjective "liberal" was broader and less questionable than it is now; but it is constitutional and liberal democracy that first England and then the United States have been representing and even bequeathing to much of the world, even now.

....

Except for a brief attempt of James II, Charles's brother in the mid-1680s, constitutional monarchy was solidly established in England by 1689, the year of the Bill of Rights and of the "Glorious Revolution" which was no revolution at all. That, unlike the brief Cromwellian rule in the 1650s, had a long effect not only in America but also in much of Europe: a not necessarily conscious but perhaps unavoidable emulation of England. Less than two hundred years later most of Europe's states were ruled by constitutional monarchies together with various parliaments. As in Britain and in the United States they had two chambers, an upper and lower (in the United States a Senate and a Congress), laws guaranteeing civil and personal rights, tolerating and respecting many kinds of religions, a considerable freedom of the presses, moving closer and closer to universal male (and eventually also female) suffrage. Together with these increasingly liberal practices, laws, and regulations (adopted even in Tsarist Russia by 1907), politics in some of Europe, whether consciously or not, by and large emulated the English (and American) two-party systems. (Of course so did the increasingly self-governing parts of the British Empire, such as Canada, Australia, New Zealand, etc.) By the eighteenth century the two great parliamentary parties in England were the Tories and the Whigs: the former conservative rather than liberal, the latter the opposite—yet during that century this division was neither categorical nor accurate: for example the greatest writer, political person and thinker upholding conservative principles was Edmund Burke, a Whig. Then, during the nineteenth century, this division between —relatively—conservatives and liberals appeared in many European states too. Unlike in England, in their parliaments there were also other parties, representing different national and social interests, among them the increasing Social Democrats. In England, too, a new and third political force was beginning to form the Labor Party (which, in the early twentieth century, would by and large replace the Liberal one). Still the conservative-liberal duality applied to politics and largely dominated it, at least until the end of the First World War.

Still we must note that as late as a century ago in Europe the term "democracy" was not yet universally approbatory or generally current. There were Democratic parties here and there, but they were generally small and usually situated somewhat to the moderate "left" of the Liberals. About this the United States was exceptional: a Democratic Party existed almost since the early Republic—the "Whig" adjective and designation, at first widely approbatory, gave way here and there to "Democratic." It was not a minority party; it elected Presidents, even though its composition and its majorities were different from the parties in England and in Europe. Democrats and Republicans: their solid domination of American politics has now remained in practice for more than one hundred and fifty years. Yet, for many reasons and sectional differences across this vast republic, to define or categorize the Democrats as largely liberal and the Republicans as largely conservative was inaccurate and misleading—less so in England, where in the nineteenth century the Conservative Liberal (the name of the first eventually replaced the Tory adjective) duality and dialogue still largely prevailed.

Meanwhile democracy was progressing throughout much of the world. But after the great cataclysm of the First World War Liberalism was not. And thus we arrive at the grave crisis of Liberal Democracy in the 1920s and 1930s, throughout much of Europe.

....

After the end of that greatest of wars, 1918, the defeat of the Central European monarchies and the "victory" of Britain and France and the United States seemed to bring a nearly universal establishment of liberal and parliamentary regimes in Europe. In 1919 in Germany too a parliamentary and liberal constitutional republic came into existence. But soon thereafter, in the 1920s and especially in the 1930s, most of the liberal regimes in Europe gave up. They were replaced by authoritarian governments, some of them dictatorships (though their parliaments, largely without much power, continued to exist at least). This happened in various places and ways, perhaps in Italy first (Mussolini), but also in Portugal and Spain and Turkey and Greece and Estonia and Latvia and Lithuania and Austria, and to considerable extents in Rumania, Hungary, Albania—and across the world, in many of the Central and South American republics. Of course the most ominous and profound of these transformations took place in Germany, where in January 1933 Adolf Hitler became its Chancellor and leader.

How did this happen? There were reasons for this development, with considerable variations from country to country. One was the rapidly weakening power and influence of the war victors Britain and France. A deeper and more widespread factor was the faded appeal of Liberalism. This had much to do with the realization that the, by and large, liberal and parliamentary regimes of the nineteenth century were less than democratic: they were composed of mostly upper-middleclass people, "bourgeois" and, even more telling, capitalists. Here and there "liberal" became an unpopular or at least unattractive adjective and word. Another reason for its rapidly declining popularity was the inefficiency and the occasional financial corruption of parliaments and governments. Yet another, probably deeper element was their seemingly deficient nationalism. Popular nationalism varied from country to country: but modern nationalism was a democratic phenomenon. Liberalism and liberal parties and liberal leaders and thinkers often under-estimated this (while many of them overestimated the importance of economics, of the existence of Economic Man, a short-sighted view of human nature). And even Liberal leaders occasionally allied themselves with the principle of "national self-determination"—which eventually led to the division and disappearance of entire empires.

So especially after the end of World War I popular support for Liberal political parties dwindled almost everywhere in Europe, even in England where the two leading parties became the Conservatives and Labor. But even more important was that the nineteenth-century Conservative-Liberal political landscape was now giving way to a new combination, that of nationalism and socialism. Of this duality the second was more universal but also less significant, since nearly everywhere in the "West" the growing welfare state was part and parcel of the advancement of their democratic societies. At the same time nationalist socialism was much more significant and potent than the international socialism dreamed up by Marx and others in the nineteenth century.

Adolf Hitler recognized this better than almost anyone else. Certainly he was no liberal—but a democrat of a kind, yes. He thought and said that even more important than the state was the people, "das Volk." National Socialism (which should never be called "Fascism," the latter having been only an Italian, and partial, phenomenon) soon had followers in Europe, indeed in many parts of the world. It impressed people, especially younger ones, who saw in it the opposite of anything aged, decadent, or reactionary: a youthful wave of the future. It led to the Second World War, for which Hitler was alone responsible, and to some of the Germans' stunning victories.

He, and his National Socialists, not the Communists,

were the most revolutionary force in the world, perhaps in the entire twentieth century. Not the Russians, not the Communists. Indeed in Russia the collapse of liberal democracy preceded even the end of the First World War. In February 1917 liberals in Russia brought about the abdication of the Tsar (already something of a constitutional monarch); but liberal democracy in Russia lasted less than a few months. The Bolsheviks or Communists put an end to it in October 1917, and then extirpated its opponents during two or more years of civil war. The Russian people, by and large, had little or no knowledge, or experience, or particular liking for those Western-type townies, liberals and bourgeois, who arrogated themselves to rule Russia in early 1917. In other places of the world—and to certain demagogues or intellectuals almost ever since—Communism meant extreme radicalism, the instrument to achieve total democracy for people. That was an illusion. What was not an illusion was the remnant power of Russia. Soon the belief in international Communism faded among the Russian masses, if indeed it ever meant anything to them. After Lenin came Stalin, another kind of a Russian Tsar. He and his Russian Tsarism survived the Second World War, turning out triumphant. Almost all of this had nothing to do with international Communism (notwithstanding the ensuing "cold war"). Until the end of the Second World War there was no Communist state elsewhere in Europe emulating Russia.

. . . .

Of this disappearance of liberalism the United States (and the nations of the British Commonwealth and the Scandinavian states and Switzerland) were the exceptions. While in the 1930s in much of Europe "liberal" was a term largely avoided, in the United States it was not a pejorative adjective—President Roosevelt employed it on occasion—while "democracy" and "democratic" were almost unquestionably approbatory, matters of fact. There were few Americans who admired the post-liberal regimes or dictatorships in Europe (the later had emulators in South and Central America, including large states such as Argentina and Brazil). In the 1930s the giant United States stood like a prevailing monument of liberal democracy, its financial "Depression" notwithstanding. Oddly or not so oddly, with all of its modernity and newness, with its idolization of Progress, the United States represented and incarnated the endurance of some things customarily traditional. We ought to recognize that in retrospect, even if few Americans thought so. The new reforms and instruments of Roosevelt's New Deal made America more and more democratic, while its liberalism too was largely unchanged.

Whether the United States would take part in a possibly coming Second World War remained at least questionable, dividing the American people, among whom men and women of British origin were no longer a majority. Still very few Americans had any sympathies for Hitler's Germany or for Mussolini's Italy—nor of course had their president, Roosevelt.

But then in September 1939 came the Second World War—which Hitler almost won. By May 1940 he had conquered and occupied almost all of the remnant democratic parliamentary states of Western Europe. Austria, Czechoslovakia, Poland, and then Denmark, Norway, Luxembourg, Holland, and Belgium were now his. He ruled more of Europe than Napoleon did. Now he had admirers and followers in Europe, indeed throughout the world, even in some of the nations that he had conquered and subdued.

By the end of May 1940 France was falling, and Britain existed in this world war alone. Then occurred a near-miracle. Even miracles have some kind of an origin, or an instrument. That instrument was Winston Churchill. He became the Prime Minister of Britain on the tenth of May—on the very day when Hitler had ordered his army to begin their great offensive in the West to subdue Holland and Belgium and France. Much of that they achieved in three weeks. Meanwhile, Churchill's Prime Ministership was shaky. Most of the Conservatives (a large majority in the English parliament then) did not like and did not quite trust him. Less than two weeks after German armies had reached the French ports on the Channel they were encircling the British army in France around Dunkirk. Now Churchill had to fight within Britain itself. For five days he had to struggle to prevail within the secret War Cabinet, five men of whom the most powerful were two Conservatives, Chamberlain and Halifax. They had no sympathies for Hitler. But Halifax, and not without reason, thought and said that the time had come to elicit something from Hitler that could satisfy him without compromising at least some of the independence of Britain. Churchill said that this would be a disaster: the first step on "a slippery slope," leading Britain to become "a slave state."

After five days he prevailed. He was fortunate, for at least one reason, which was that Hitler was not quite sure of himself. The British and French soldiers (the latter about one-fourth of the British), surrounded at Dunkirk, were lifted off to Britain in a difficult seaborne operation. Hitler was not entirely disappointed with that. He thought that, sooner rather than later, the British political elite would replace Churchill. This corresponded with something that even now few people realize. He did not aim to conquer the world, or even most of it. He wanted to rule Europe. That Churchill knew. Aware of centuries of English history he understood that England should, or could, not accept this. And there was another reason for his sturdiness. He had corresponded with Roosevelt for some months now. He knew that Roosevelt and the United States would support Britain—though when? And how? Churchill's third reason ("third" not merely in order of its importance) was his trust in the British people. He sensed that his words would inspire them, because of his patriotism but also because of his truthfulness. In his very first speech after he had become Prime Minister: "I have nothing to offer but blood, toil, sweat and tears." When he had summoned his entire cabinet after his dreadful five days he said: "whatever happens at Dunkirk we will fight on." His words brought an instant eruption of relief and support.

If Halifax had had his way—which, I repeat, was not altogether unreasonable—Hitler may have won his war. Had in 1940 Herbert Hoover or someone like him been President of the United States Britain could not have prevailed and Hitler may have won his war. What would have come if Hitler had won? We cannot tell. But his rule would have lasted a long time: more pervasive, much newer, more lasting than those of other dictatorships, such as Communism. So, in 1940, perhaps for the last time in the history of the West, indeed perhaps in the history of the white race, England was the protagonist of what remained of liberty: the first and last bastion of freedom.

Even after 1940 England was the key to Hitler's thinking about the war. In 1941 he invaded Russia not because of his wish to destroy Communism or to win more lands for Germany in the East: but because—as he himself said and admitted to his generals on many occasions—Russia was Churchill's (and Roosevelt's) last hope. Once he had conquered Russia what could they do? They, and their peoples, would be forced to seek a compromise peace with him. He may allow the British to keep some of their world-wide empire; but Europe must be largely under German rule. Well, this did not happen. But also consider how a Germany of 80 million people fought a war with the British and American and Russian empires, perhaps 500 million people in toto; and for them to conquer Germany took almost six years. The British alone could not do it. Neither could Russia. Perhaps the British and the Americans together could, but neither could do it by themselves. The result was that Russia ended up ruling Eastern Europe; and for Britain and America Western Europe, that better half of Europe, was better than none.

They saved the Western world. America's war was awesome. It won on both sides of the globe, in the Atlantic and in the Pacific. The Japanese thought that they could fight the United States to achieve a peace similar to what they had exacted from Russia fifty years before: but that would not be. "The West"—with the United States as its new protagonist—had won.

. . . .

For a while—and to very many people—it seemed that after the Second World War the great struggle across the world, and perhaps even a Third World War, was due between "Democracy" and "Communism," represented and incarnated by the now two superpowers, the United States and Russia. In some ways this was so, but in other ways it was not. The end of the Second World War, unlike the end of the First, did not bring about widespread revolutions. Indeed in 1945 and after, liberal democracy was not only restored in Western Europe (including in Western Germany) but many of its practices and features spread on and on. Even in the Russia-occupied states in Eastern Europe there were largely free elections, various parties, some freedom of the press, etc., until about 1947 when Stalin and the rulers of Russia imposed Communist and one-party totalitarian rules, sealing off their Eastern European domains with an "iron curtain" from the West. Far away from the Soviet Union there were nationalist Communist dictatorships in odd places of the globe, from Cuba to China, etc., etc., but nowhere in the West. After a few decades the appeal of more and more personal freedoms and communication then began to affect and change the dictatorial rules of the remnant "Communist" regimes too.

The United States was now, and still is, the only Superpower in the world—again a champion and protector of democracies and many of their liberal features. What was, and remains, interesting is the rapid decline of the largely popular impression that American liberals were not sufficiently nationalist, that they were more tolerant of Communists and of Communism than were most Americans—especially when Communism seemed to be the only great world-wide threat that the United States now faced. As late as in 1951 Senator Robert A. Taft, leader of the former isolationist wing of the Republican Party, still called himself "an old-fashioned liberal." That usage disappeared among the Republicans who elected presidents in the 1950s, at the end of which even President Eisenhower called himself a "conservative." That adjective, almost unemployed and untouched by American politicians earlier, had become an approved word among Republicans. Their party and their followers had, by and large, become nationalists and interventionists and even populists. One example was the meteoric rise of the popularity of Senator Joseph McCarthy, who was a populist nationalist and not only anti-liberal but also a threat to some of the practices and traditions of American liberalism. McCarthy was declining by 1956, but in the same year the Republican Party's platform called for the establishment of American army and air bases "all around the world." (In 1956 there were about 150 of these outside of the United States; thirty years later more than 900.) And now? It was not until the second decade of the twenty-first century that more and more Americans began to question the intervention of American armies and naval and air forces in the so-called Middle East.

Whatever the decline of "liberal" may have meant in Western Europe, an international institution of something like a European Union in 1947–48 was significant. Its shortcoming was, and still is that it has been economic and financial rather than political, believing (another old liberal conviction) that economic changes will bring about political ones. This is not what happened. The European Union has been impotent and unwilling, as when in 1991 it left a murderous civil war in Yugoslavia to be suppressed by American air power, without any "European" intervention at all. Perhaps—perhaps—the European Union will, after some time, develop into something truly international. That will not come for a long time. Perhaps—perhaps—"Europe" may become like Switzerland, the Switzerland on the globe. But history seldom repeats itself.

Democracy brought about nationalism; but nationalism is, more than often, a threat to liberalism. It may be significant that many of the Eastern European nations free from Russian or Communist rule in 1989 have tended in a populist and nationalist direction, at times hostile to the

European Union. "Democrat" and "democracy" are positive designations (and even titles of publications) in most of these nations, while "liberal" is customarily avoided, and at times even accusatory, yet liberal-democratic customs and traditions—not only universal suffrage but free elections, parliaments, multiple parties, freedom of publications and of communications still largely prevails, desired as they are by many people. Whatever its designation, Liberal Democracy, a creation of the West, still prevails—here and there, across the globe. It will not last forever—nothing does—but it will for a long time. Recognizing this is what I call a faint reason for optimism, even when so many ominous cracks of civilization at the end of an age are manifest.

An intelligent Frenchman wrote recently a book on colors, now one big one on "Green." I ordered it but have not received it yet. He mentions Goethe's "Theory of Colors" but without much emphasis on it, I think. Yet that was a very important work, though disregarded for too long: that color has three ingredients: its chemical substance, its very dependence on contrast, *and* the act of seeing. (We are at the center of the universe, etc.) I wrote this quatrain:

How I love, I love all that looks green
Not that I know what one color does mean
All I know is that I can see it so
Ergo sum benedicatus Domino.